Health and Climate Change

Health and the Environment Series
Edited by Erik Millstone

Health and Climate Change

Modelling the Impacts of Global Warming and Ozone Depletion

Pim Martens

First published in the UK in 1998 by
Earthscan Publications Ltd

This edition published 2013 by Earthscan

For a full list of publications, please contact

Earthscan
2 Park Square, Milton Park, Abingdon, Oxfordshire OX14 4RN
Simultaneously published in the USA and Canada by Earthscan
711 Third Avenue, New York, NY 10017

First issued in paperback 2014

Earthscan is an imprint of the Taylor & Francis Group, an informa business

A catalogue record for this book is available from the British Library

ISBN 13: 978-1-85383-523-0 (hbk)
ISBN 13: 978-0-415-84880-0 (pbk)

Typesetting and page design by Pim Martens

Cover design by Andrew Corbett

Contents

List of Acronyms and Abbreviations

ABS	Australian Bureau of Statistics
BAF	Biological amplification factor
BCC	Basal cell carcinoma
CBS	Central Bureau of Statistics
CDC	Centers for Disease Control
CFC	Chlorofluorcarbon
CHD	Coronary heart disease
CVD	Cardiovascular disease
DDT	Dichloro-diphenyl-trichloro-ethane
DHF	Dengue haemorrhagic fever
DNA	Deoxyribonucleic acid
DSS	Dengue shock syndrome
EIP	Extrinsic incubation period
EP	Epidemic potential
EPA	Environmental Protection Agency
GCM	General Circulation Model
HBI	Human blood index
HCFC	Hydrochlorofluorcarbon
IHD	Ischaemic heart disease
IPCC	Intergovernmental Panel on Climate Change
MIASMA	Modelling framework for the health Impact Assessment of Man-induced Atmospheric changes
MSC	Melanoma skin cancer
NMSC	Non–melanoma skin cancer
NOAA	National Oceanic and Atmospheric Administration
RD	Respiratory disease
RIVM	Dutch National Institute of Public Health and the Environment
SCC	Squamous cell carcinoma
TARGETS	Tool to Assess Regional and Global Environmental and health Targets for Sustainability
TOMS	Total ozone mapping spectrometer
UN	United Nations
UNEP	United Nations Environment Programme
UV	Ultraviolet

VC   Vectorial capacity
WHO   World Health Organisation
WMO   World Meteorological Organisation

List of Figures

List of Tables

List of Boxes

Foreword

Early in the 1990s a new and large environmental health hazard appeared on the horizon. We perceived that no longer were we just polluting local environments, depleting local resources stocks, or causing regional acid rain. We were, as well, beginning to change some of the great biophysical structures and processes of the planet. As the scale of human impact on the world's environment continues to escalate, we face a qualitative shift in the types of environmental disturbances that might result.

The bread-and-butter task of epidemiologists is to describe and explain variations in rates of disease or death between different populations, groups, or subsets of individuals sharing some common feature. That research usually refers to either the present or the recent past. It is empirical, often hypothesis-testing, and can be addressed with statistical models that describe the observed relationship. Such relationships, once determined, can also be used to forecast the future health outcomes of present circumstances - for example, tomorrow's AIDS incidence as a function of today's HIV seroprevalence, or tomorrow's lung cancer mortality as a function of today's cigarette smoking behaviours. But a move into the "future provisional" tense confronts us with a less familiar task. First we must ask: What are the plausible scenarios of future environmental circumstances (and future social arrangements)? Then, how might we best model the health impacts of those scenarios, especially if they entail conditions that go outside the boundaries of our documented experience? And how do we handle the unavoidable uncertainties?

This futuristic research question has emerged as an increasingly important and urgent task for public health science. In the early 1990s, several forward-thinking groups of scientists began to colonise this new ecological niche in the research domain. They began developing various approaches to the integrated assessment of scenario-based health risks. Pim Martens and his mentors and colleagues in The Netherlands have been early leaders in this challenging field. Their teamwork and their broadly-based, interdisciplinary, approach has yielded valuable progress in the mathematical modelling of potential health consequences of current and anticipated global changes.

Combining a background in mathematics, systems modelling and environmental health sciences, Pim Martens has carried this work forward in relation to three much-discussed potential health consequences of global change:

shifts in the geography of vector-borne infectious diseases due to the anticipated changes in regional climatic conditions; alterations in exposure to thermal stress in urban populations under conditions of climate change; and increases in the incidence of skin cancer in selected populations in response to continuing losses of stratospheric ozone over the coming decade or so, followed by its gradual recovery to "pre-CFC" levels. Each of these three topics is treated in detail in this book.

Of particular interest, each of those three problem areas has it own qualitatively distinct features. The book therefore provides a sturdy foundation for thinking about how best to tackle a varied spectrum of population health hazards posed by different aspects and combinations of global change processes. The mathematical models used, of course, have their limitations and these are explicitly discussed. Much of this early, ground-breaking, work has been of a highly aggregated kind, some of it frankly global. However, Martens has also begun to focus down on regional impacts, as for example in the modelling of climate-change impacts on malaria transmissibility in Zimbabwe. The book also explores the need to move beyond systems-based models to those that recognise the changeability of complex adaptive systems, that is, systems that evolve in response to changing circumstances. For example, malaria transmissibility is affected not just by changes in mosquito-parasite system responses to altered temperature and rainfall, but by factors such as whether populations move or become more immune, and whether mosquito populations become pesticide resistant and the malarial parasite develops drug resistance. Complexity theory, genetic algorithms, artificial neural networks and so on - all of which have themselves evolved recently in response to our changing perceptions of the integrated dynamic world around us - can facilitate this extension of our health-impact modelling techniques.

The work presented here also goes the extra mile. In the final chapter, Martens estimates the attributable population burdens of disease or mortality that are likely to result from these aspects of global change. Too often in epidemiology we have rested on our laurels after identifying a relationship or quantifying the risks due to specified exposure levels. There is always more to do. Communities and their policy-makers want to know something about the overall magnitude, something about the relative importance, of the predicted health impact. It is therefore heartening to see here the results of this mathematical modelling being presented in policy-relevant terms.

In a decade's time we will have inevitably become better at carrying out scenario-based modelling. We will then look back and see that the research conducted in the mid-1990s, well exemplified by this book, was fundamentally important in laying foundations and in alerting a wider constituency of public health scientists to this unprecedentedly large and challenging task. As our environmental impacts escalate, it is no longer appropriate for epidemiologists to quietly tidy up in society's wake. We must help our societies, and humankind at large, to see the possible future consequences of today's actions.

Tony McMichael

Professor of Epidemiology
Department of Epidemiology and Population Health
London School of Hygiene and Tropical Medicine

Preface

This book represents the culmination of four years of research at the Department of Mathematics, Maastricht University, and the research programme 'Global Dynamics and Sustainable Development' at the Dutch National Institute of Public Health and the Environment (RIVM) in Bilthoven. It deals with an eco-epidemiological modelling framework, MIASMA, which consists of computer simulation models dealing with the effects of climate change on vector-borne diseases and thermal-related mortality, and the effects of ozone depletion on skin cancer incidence. The models are meant to increase our insights into the underlying processes of climate change, ozone depletion and human health, and are intended to stimulate and contribute to the ongoing discussion on the development of methods in the analysis of the interactions between atmospheric changes, ecosystems and human health.

Most of the material presented in this book is based on my PhD dissertation at Maastricht University on *Health Impacts of Climate Change and Ozone Depletion: An Eco-Epidemiological Modelling Approach* (Martens, 1997a). During my PhD research I have been fortunate to be surrounded by many colleagues who have supported me and provided me with critical but stimulating comments on my research activities. First of all, I owe many thanks to my Professors, Jan Rotmans and Koos Vrieze, for their support and many interesting discussions through the years; discussions which undoubtedly have had a substantial influence on my academic development. I would like to thank Tony McMichael for his early interest in my research, and for the time he took to discuss my work during my visits to London. His lucid discussions on methods for assessing the impacts of environmental change on human health contributed greatly to the writing of Chapter 2.

Next, I am grateful to many people for their contributions to my work on vector-borne diseases, described in Chapters 3 and 4. Special thanks to Dana Focks, Theo Jetten, Steve Lindsay, Louis Niessen, and Jonathan Patz for sharing their knowledge and time. I thank Marco Janssen for his stimulating and enthusiastic discussions on complex adaptive systems and for his cooperation in applying evolutionary modelling techniques on my malaria model, described in Chapter 4. I would like to thank Michel den Elzen and Harry Slaper for their substantial contribution to the analysis of ozone depletion and skin cancer incidence, described in Chapter 6. Thanks are also due to Petra Koken and Ben Willems. Without their help the implementation of the skin cancer model for Australia could never have been done. Furthermore, thanks to all other colleagues at the Department of Mathematics, Maastricht University, and at the RIVM, for their support, and pleasant and stimulating conversations.

As a modeller, I know how important a good back-up is. Therefore, last but not least, I would like to thank Petra and my parents and family, who showed continued interest and support throughout the years. I dedicate this book to them.

Pim Martens
Maastricht
October 1997

Chapter 1

INTRODUCTION

THE ISSUE

The health of a population, if it is to be maintained in a 'sustainable state' (King, 1990), requires the continued support of clean air, safe water, adequate food, tolerable temperature, stable climate, protection from solar ultraviolet radiation, and high levels of biodiversity. Socio-economic changes and health interventions have improved public health in recent decades, although there are still many disparities in fulfilled health potential on the global level and amenable morbidity and premature mortality continue to exist (World Health Organisation (WHO), 1995a). However, as a counter-effect of economic development, health impairments have started to occur as the result of deteriorating global environmental conditions.

Global environmental change is a general umbrella term for a whole range of mutually dependent global environmental problems attributable to human activities. They include acidification, eutrophication, deforestation, land degradation and desertification, loss of biodiversity and depletion of fresh water supplies. Major global environmental changes that can be expected to have a significant health effect include climate change and ozone depletion (McMichael et al., 1996). Human-induced climate change and stratospheric ozone depletion are now threatening the sustainability of human development on the planet, because they threaten the ecological support systems on which human life depends (McMichael, 1993), as well as human health and wellbeing, the continuing improvement of which should be the very goal of the development process itself. King (1990) has pointed to the widespread agreement that numerous developing countries are 'demographically trapped', in that communities have exceeded or are projected to exceed the carrying capacity of their local ecosystems, their ability to migrate, and the ability of the economies to produce goods and services in exchange for food and necessities. These gross failures in sustainable development are marked by the health patterns associated with infectious diseases, malnutrition and starvation (partly relieved by food aid). Climate change and depletion of the ozone layer could inflict severe additional stress on such already overburdened ecosystems.

Climate Change and Ozone Depletion

Greenhouse gases allow incoming solar radiation to pass through the atmosphere but trap the re-radiated long-wave radiation from the Earth's surface (see Figure 1.1). Since the Industrial Revolution, human activities have increased the atmospheric concentrations of the greenhouse gases, leading to the enhanced greenhouse effect. The main anthropogenic greenhouse gases are carbon dioxide (CO_2), methane (CH_4), nitrous oxide (N_2O), and chlorofluorocarbons (CFCs).

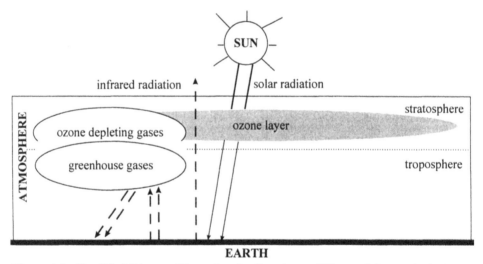

Figure 1.1: Simplified Diagram Illustrating the Greenhouse Effect and Stratospheric Ozone Depletion

The concentration of CO_2 has increased by around 25 per cent primarily due to emissions from fossil fuel burning and deforestation. CH_4 has more than doubled in concentration since 1750. Sources of CH_4 are less certain than those of CO_2 but include rice paddies, animal and domestic waste, coalmining and venting of natural gas. The atmospheric concentration of N_2O has also been growing since the mid-18th century, and its sources include nylon production, three-way catalytic converters in cars, and, possibly, agriculture. CFCs, used in refrigerators, air conditioners, and foam insulation, are not only involved in greenhouse warming in the troposphere, but also in the depletion of the ozone layer in the stratosphere. Recently, evidence has accumulated that sulphate aerosols (products of sulphur dioxide) may dissipate solar radiation and thus prevent it from reaching the Earth's surface, thereby masking the enhanced greenhouse effect over some parts of the Earth (Charlson & Wigley, 1994; Intergovernmental Panel on Climate Change (IPCC), 1996). Changes in the concentration of greenhouse gases and aerosols, taken together, are projected to lead to regional and global changes in climate and climate-related parameters such as temperature and precipitation. Although the reliability of regional projections is still poor, and the degree to which climate variability may change is uncertain, climate models project an increase in global mean surface temperature of about 1°C–3.5°C by 2100 (IPCC, 1996).

Ozone (O_3) is an atmospheric trace gas, 90 per cent of which is distributed in the stratosphere, mostly between altitudes of 15–25 km. The destruction of the stratospheric ozone layer is largely attributed to reactive chlorine, liberated from mainly CFCs under specific meteorological conditions in the stratosphere (sunlight and stratospheric cloud formation). In the stratosphere ozone acts like a protective shield, preventing much of the sun's ultraviolet radiation (UV), especially UV with shorter wavelengths, from reaching the Earth. Over the past few decades, stratospheric ozone concentrations have fallen globally, especially during winter and springtime at higher latitudes, and most markedly over the Antarctic. Over approximately the period 1980–1990 ozone depletion at northern latitudes of 30–60°N has been 6 per cent during winter and spring, and 3 per cent during summer and autumn. In the southern hemispheric, cumulative ozone depletion has amounted to 5 per cent per decade since 1980 (United Nations Environment Programme (UNEP), 1994). As a result of this decrease in the thickness of the ozone layer, more UV-B radiation is expected to reach the earth.

The problems of climate change and ozone depletion are interrelated, not only by virtue of their common sources, but also as a consequence of the numerous interrelations between them. Depletion of ozone in the lower stratosphere could cause a reduction in the radiative forcing, which could offset a fraction of the global warming attributed to the increases in the abundance of greenhouse gases (Ramaswamy et al., 1992; World Meteorological Organisation (WMO), 1992). The increases of greenhouse gases in the atmosphere and the resulting tropospheric warming and stratospheric cooling will affect the ozone concentrations in the stratosphere: the decrease in the stratospheric temperature will slow down the rate of ozone destruction induced by chemical reactions. An indirect impact of rises in UV radiation associated with global warming is the impairment of the primary production of phytoplankton, which will limit the oceans' role in acting as a sink for CO_2. This will lead to an increase in atmospheric CO_2 concentrations (den Elzen, 1993).

Nevertheless, the essential difference between greenhouse accumulation and stratospheric ozone depletion should be borne in mind. Greenhouse gas accumulation increases the effect of radiative forcing on climate, while stratospheric ozone depletion by chlorine radicals leads to increased UV radiation at ground level. These two distinct phenomena are thus members of a wider-ranging family of global atmospheric changes.

Health Impacts

Increased levels of UV radiation due to ozone depletion may have serious consequences for living organisms. Adverse impacts of UV-B have been reported on terrestrial plant growth and photosynthesis. Increased UV-B has also been shown to have a negative influence on aquatic organisms, especially small ones such as phytoplankton, zooplankton, larval crabs and shrimps, and juvenile fish. Since many of these organisms are at the base of the marine food chain, increased UV-B may seriously affect aquatic ecosystems (for more details on these effects see, e.g., UNEP, 1994; WHO, 1994). Furthermore, increased UV-B radiation affects tropospheric air quality and may cause damage to materials such as wood, plastics and rubber.

Climate changes are also likely to be associated with a multitude of effects: climate change will shift the composition and geographic distribution of many ecosystems (e.g. forests, deserts, coastal systems) as individual species respond to changed climatic conditions, with likely reductions in biological diversity; agricultural yields may also be affected. Climate changes will lead to an intensification of the global hydrological cycle and may have impacts on regional water resources. Additionally, climate change and the resulting sea-level rise can have a number of negative effects on energy, industry and transportation infrastructure, human settlements, and tourism (for more details see IPCC, 1996).

Only recently has attention been paid to the possible consequences of these global atmospheric changes for human health (e.g. WHO, 1990; Haines & Fuchs, 1991; Doll, 1992; McMichael, 1993, 1996; McMichael et al., 1996). Figure 1.2 summarises some of the important potential effects of climate change and ozone depletion upon human population health. Broadly speaking, the various potential health effects of global climate change upon human health can be divided into direct and indirect effects, according to whether they occur predominantly via the impacts of climate variables upon human biology, or are mediated by climate-induced changes in other biological and biogeochemical systems.

In healthy individuals, an efficient regulatory heat system enables the body to cope effectively with thermal stress. Temperatures exceeding comfortable limits, both in the cold and warm range, substantially increase the risk of (predominantly cardiopulmonary) illness and deaths. Directly, an increase in mean summer and winter temperatures would mean a shift of these thermal-related diseases and deaths. An increased frequency or severity of heatwaves will also have a strong impact on these diseases. If extreme weather events (droughts, floods, storms, etc.)

were to occur more frequently, increases in rates of deaths, injury, infectious disease and psychological disorder would result.

One of the major indirect impacts of global climate change upon human health could occur via effects upon cereal crop production. Cereal grains account for around two-thirds of all foodstuffs consumed by humans. These impacts would occur via the effects of variations in temperature and moisture upon germination, growth, and photosynthesis, as well as via indirect effects upon plant diseases, predator-pest relationships, and supplies of irrigation water. Although matters are still uncertain, it is likely that tropical regions will be adversely affected (Rosenzweig *et al.*, 1993), and, in such increasingly populous and often poor countries, any apparent decline in agricultural productivity during the next century could have significant public health consequences. A further important indirect effect on human health may well prove to be a change in the transmission of vector-borne diseases (Patz *et al.*, 1996). Temperature and precipitation changes might influence the behaviour and geographical distribution of vectors, and thus change the incidence of vector-borne diseases, which are major causes of morbidity and mortality in most tropical countries. Increases in non-vector-borne infectious diseases, such as cholera, salmonellosis, and other food- and water-related infectious diseases could also occur, particularly in (sub)tropical regions, due to climatic impacts on water distribution, temperature and the proliferation of micro-organisms.

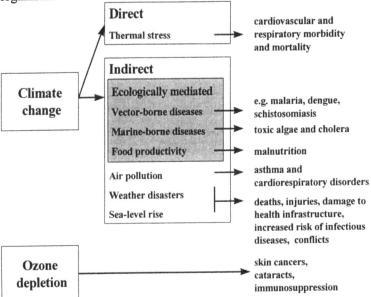

Figure 1.2: Health Impacts due to Climatic Changes and Ozone Layer Depletion (Source: Patz & Balbus (1996))

Table 1.1: Summary of Known Effects and Uncertainties Regarding Health Impacts of Climate Change and Ozone Depletion

Health effect	Known effects	Uncertainties
Thermal stress	* Mortality (especially cardiopulmonary) increases with cold and warm temperatures * Older age groups and people with underlying organic diseases are particularly vulnerable * Mortality increases sharply during heatwaves	* The balance between cold- and heat-related mortality changes * The extent to which heatwaves take their toll among terminal patients * The role of acclimatisation of people to warmer climates
Vector-borne diseases	* Climate conditions (particularly temperature) necessary for some vectors to thrive and for the micro-organisms to multiply within the vectors are relatively well known	* Indirect effects of climate change on vector-borne diseases, such as changes in vegetation, agriculture, sea-level rise, migration, etc. * Effects of socio-economic development, resistance development, etc.
Water-/ food-borne diseases	* Survival of disease organisms (and insects which may spread them) is related to temperature * Water-borne diseases most likely to occur in communities with poor water supply and sanitation * Climate conditions affect water availability * Increased rainfall affects transport of disease organisms	* For many organisms the exact ambient conditions in which they survive and are transmitted are not known * Interaction with malnutrition is not well understood
Food production	* Temperature, precipitation, solar radiation and CO_2 are important for crop production * Crop failure may lead to malnutrition * Undernourishment may increase susceptibility to infectious diseases	* Variations in crop yield due to climate change are poorly understood * Effects of climate on weeds, insects and plant diseases are not well known * Interaction between nutritional status and diseases is poorly understood
Skin cancer	* Skin cancer incidence is related to UV exposure * Ageing increases the risk of skin cancer	* Dose-response relationship between UV radiation and skin cancer, especially basal cell carcinoma and melanoma skin cancer, is not completely clear
Cataracts	* UV radiation damages the eye, more particularly the lens * Different types of cataracts will react differently to changes in UV radiation * Aetiology of cataracts is associated with age, diabetes, malnutrition, heavy smoking, hypertension, renal failure, high alcohol consumption, and excessive heat	* Dose-response relationship between UV radiation and cataracts is not well known * Interactions with other determinants of cataracts are not always clear
Immune suppression	* UV suppresses immune systems in animal models, and may adversely affect various infections * In man, serial UV irradiation may cause proper immunisation to fail * UV-induced immunosuppression appears to be a risk factor for skin carcinomas	* Interaction between immunosuppression and infectious disease incidence * Effect of immunosuppression on vaccination efficacy

Many health impacts could also result from deterioration in physical, social and economic circumstances caused by rising sea levels, climate-related shortages of natural resources (e.g. fresh water), and impacts of climate change on population mobility and settlement. Conflicts may arise over decreasing environmental resources. The climate change process is associated with air pollution, since fossil fuel combustion produces various air pollutants. Furthermore, higher temperatures would enhance the production of various secondary air pollutants (e.g. ozone and particulates). As a consequence, there would be an increase in the frequency of allergic and cardiorespiratory disorders and deaths caused by these air pollutants.

If a long-term increase of UV-B radiation due to stratospheric ozone depletion occurs, melanoma skin cancer (MSC) and non-melanoma skin cancer (NMSC) will increase, people with lightly pigmented skins being most susceptible. The incidence of various diseases of the eye, particularly pterygium and cataracts, is also likely to increase. There is less certainty about whether damage to the human immune system (both local and systemic) may occur, leading to increased vulnerability to infectious diseases. A potentially more important, indirect effect of increased UV-B levels reaching the Earth could be the UV-B-induced impairment of photosynthesis on land (food crops) and in the sea (phytoplankton), reducing the world's food production.

The potential impacts of climate change and ozone depletion cannot be separated. The health effects are very likely to be synergistic and may be cumulative in vulnerable populations, which is illustrated by the following issues. (i) The paramount health problem in the world appears to be that of nutrition and infectious diseases and their interrelationships: malnutrition may exacerbate risks of morbidity and mortality and the increased prevalence of infectious disease is thought to increase malnutrition. An anthropogenic climate change is likely to influence both. (ii) The human immune system is likely to be adversely affected by increased UV-B radiation. This would reduce protection against infectious and fungal diseases, and could reduce vaccination efficiency in immunologically marginal (i.e. undernourished and chronically infected) populations, the same populations that are likely to be most affected by climate change-induced reductions in agricultural production, leading to the gradual (re-)appearance of infectious diseases. (iii) The prevalence of cataracts is often much higher among elderly, malnourished persons in poor countries, where micronutrient deficiency may contribute to cataract formation (Harding, 1992). Thus, besides being an effect of increasing UV-B levels, cataract prevalence may be influenced by changes in the availability of food.

However, many uncertainties remain, not only regarding these health impacts, but also related to scenarios of climate change and ozone depletion. Table 1.1 gives a survey of what is known and the uncertainties in relation to the health impacts discussed above.

SCOPE AND OBJECTIVES

The potential consequences of the atmospheric changes for human health, discussed above, have only recently been placed on the international agenda. Although the adverse health consequences of stratospheric ozone depletion were discussed at several scientific conferences in the mid-1980s, less attention was given to the potential health impacts of climatic changes caused by greenhouse gas emissions. In 1990 the WHO published a report by an expert panel on the *Potential Health Effects of Climate Change* (WHO, 1990). This report paid particular attention to the health impacts of heat stress, air pollution, malnutrition due to impaired food productivity, the potential change in vector-borne disease distribution, and flooding. An assessment of the impacts of ozone depletion has been published by the UNEP (1991, 1994).

Recently, the health implications of the effects of global warming and stratospheric ozone depletion have been assessed by a Task Group convened by the WHO, the WMO and the UNEP (McMichael *et al.*, 1996); the IPCC for the first time included a separate chapter on the health impacts of climate change and ozone depletion in their assessment report (McMichael, 1996).

Despite the increasing awareness of these effects there is certainly a need for a more comprehensive and quantitative evaluation of the impact of global climatic changes and stratospheric ozone depletion on human health, which constitute, on aggregate, a more fundamental hazard to human population health than any that have gone before. They also present an array of major scientific challenges, both conceptually and technically, in the assessment of health impacts. Central to this challenge to public health science is the need to move from a reliance on empirical data describing the past to the employment of anticipatory thinking and the mathematical modelling of potential *future* impacts. A major task for public health science is to provide policy-makers, and their public constituency, with a clearer description of the anticipated future health impacts of global atmospheric change. New techniques and approaches will be needed to deal with the substantial uncertainties that inevitably surround these estimates.

Initial projections of mankind's vulnerability to global climate change have focused on the changes in vulnerability of such natural and social systems as forests, food crops, fisheries, coastal areas and physical structures, to an increased frequency of extreme weather events. The effects of these catastrophic events would undoubtedly have an impact on the health of human populations (e.g. environmental health infrastructural damage is to be anticipated from weather disasters and sea-level rise, aggravated by climate-forced human migration), and many of the health implications of weather disasters are well known. However, in relation to the more *primary* health impacts there has been little development of formal mathematical models for even relatively straightforward relationships, such as those between a change in ambient temperature and changes in mortality due to thermal stress. To fill this lacuna, this book presents an eco-epidemiological modelling framework, MIASMA (Modelling framework for the health Impact ASsessment of Man-induced Atmospheric changes). The framework has been developed at Maastricht University, in close cooperation with the Dutch National Institute of Public Health and the Environment (RIVM). A global integrated assessment model entitled TARGETS (Tool to Assess Regional and Global Environmental and health Targets for Sustainability) has been developed within the research programme 'Global Dynamics and Sustainable Development' at the RIVM (of which the author has been a member since 1993). TARGETS is intended to perform an analysis and assessment on a global scale of social and economic processes, biophysical processes, and effects on ecosystems and humans from an integrated systems dynamic perspective (Rotmans & de Vries, 1997). MIASMA can be considered as a spin-off from the TARGETS model, focusing on potential health impacts of climate change and ozone depletion.

An important objective of this book is to provide a better understanding of the dynamics underlying the health impacts of climate change and ozone depletion, and the sensitivities and uncertainties surrounding these impact estimates. In this context, the instructive and educational value of models presented can strongly increase our awareness of the potential health impacts of global atmospheric changes. Another goal of the modelling framework is to help identify gaps in our knowledge of the processes underlying the impacts studied, which may help to set the agenda for further research. Furthermore, the models and techniques discussed in this book are meant to contribute to the ongoing discussion and development of methods in the analyses of atmospheric change, ecosystems and health relationships.

For practical reasons (such as time constraints, lack of data or scientific evidence) only a few of the atmospheric change-related (primary) health impacts are discussed, including the effect of climate change on vector-borne diseases (malaria, dengue and schistosomiasis) and thermal-related mortality (e.g. cardiovascular and respiratory mortality), and the effects of increasing UV levels due to ozone depletion on skin cancer rates. Quantitative estimates exist for some of the impacts not considered in this book; for example Rosenzweig et al. (1993) assessed the additional number of people at risk of malnourishment as a result of climatic changes to be 40 million–300 million in the year 2060, and Kalkstein (1993) estimated a several-fold increase in deaths due to heatwave-related mortality towards 2050. For other impacts, such as immunosuppression due to increased UV radiation, empirical evidence upon which to base a risk assessment remains inadequate. Furthermore, many other important influences on population health are changing and will interact with the effect of climate change and ozone depletion. For example, new vaccines are being developed and safe drinking water and sanitation are becoming available to an increasing number of people in developing countries (WHO, 1996), which will undoubtedly affect the transmission of (vector-borne) infectious diseases. On the other hand, increasing rates of tobacco use will cause a rise in diseases and deaths from cigarette smoking, and infectious diseases are emerging due to, among other things, increases in drug and pesticide resistance. However, neither the scope of this book nor the available scientific evidence allows a comprehensive consideration of all potential health impacts of atmospheric changes.

OUTLINE

The outline of this book is as follows: Chapter 2 discusses the main differences between conventional epidemiological research methods and a systems-based health impact assessment of global atmospheric changes. Furthermore, this chapter discusses the rationale behind MIASMA. Chapter 3 studies the effect of climatic changes on vector-borne diseases, particularly malaria, schistosomiasis and dengue. Chapter 4 introduces an extension of the malaria model discussed in Chapter 3, including a genetic algorithm to simulate the effect of the use of insecticides and drugs to control malaria, and the development of resistance of the malaria mosquito and parasite to these control measures. The direct impact of climate change through changes in thermal stress-related mortality is discussed in Chapter 5. An assessment of the effect of stratospheric ozone depletion on skin cancer rates for The Netherlands and Australia is given in Chapter 6. Finally, Chapter 7 concludes the book and contains a discussion of possible future lines of research on the assessment of the health impacts of climate change and ozone depletion.

Chapter 2

AN ECO-EPIDEMIOLOGICAL MODELLING APPROACH

INTRODUCTION

Although there is a growing awareness that climate change and ozone layer depletion pose problems for human society, as discussed in the previous chapter, the consequences for human wellbeing of these global atmospheric changes has been less well anticipated. The health hazards posed by these changes would, to a large extent, differ qualitatively from those due to the direct-acting toxicity of local environmental pollutants. Hence, the advent of these global atmospheric changes is markedly broadening the scope of contemporary environmental health problems (McMichael, 1993; McMichael & Martens, 1995).

This chapter first discusses the main differences between conventional epidemiology and an eco-epidemiological modelling analysis. Next, MIASMA is discussed, and some key issues in the development of integrated mathematical models in the health impact assessment of climate change and ozone depletion are reviewed.

LIMITATIONS OF CONVENTIONAL EPIDEMIOLOGY

Environmental epidemiology generally refers to the influence of environmental factors on human health that are outside the immediate control of the individual (Rothman, 1993). Exposures of interest to environmental epidemiologists include air pollution and water pollution, and occupational exposure to physical and chemical agents. The formal study of the risks associated with these health hazards has contributed much to the evolution of modern, quantitative, epidemiological research methods. The risks to health are in general easily understood, since they entail evident causal effects and linear (or otherwise orderly) relationships, and act via direct toxicological damage to organ systems or metabolic pathways.

Conventional (environmental) epidemiological research thus relies on empirical data that describe the prior real-world experiences of human populations; the risks to health associated with specified factors are then estimated, by comparing rates or proportions or by fitting statistical models to the data.

By contrast, most impacts of climate change and ozone depletion would not occur via the familiar toxicological mechanisms that mediate the effects of localised exposure to environmental pollutants. Rather, they would arise via more complex processes that result from disturbances of natural global biogeochemical cycles, and the scale of these effects would primarily apply to populations or communities, rather than to individuals, for whom a small risk is increased, while nevertheless remaining small.

Furthermore, assessment of the future impacts of global atmospheric changes concerns the potential effects of an anticipated exposure, i.e. assessment of future (scenario-based) possibilities rather than the estimation of risks from past realities. The key point is that there is a fundamental difference between a data-based, explanatory approach and a scenario-based, descriptive approach. The latter uses existing information about climate and UV-related factors, such as infectious agents and skin cancer rates, to project how a change in such factors would affect populations.

Some health impacts of global atmospheric changes can be estimated by reasonable 'extrapolation' of relatively simple cause-effect models. For example, a change in ambient temperature is expected to change the number of thermal-related deaths (see Chapter 5). However, this may not be appropriate if the health risk concerned is linked to an ecological phenomenon or entity. Infectious diseases are the most obvious example of a category of health problem with complex, climate-related, ecologically based dynamics (see Chapters 3 and 4).

Table 2.1: Main Differences between Conventional Epidemiology and Eco-Epidemiology

Conventional epidemiology	Eco-epidemiology
Toxicological	Ecological
Estimation of risk from past realities	Assessment of future health risks
Short time horizon	Long time horizon
Estimation of more local risks	Estimation of global and regional risks
Statistical models	Mathematical models
Static cause-effect	System-dynamic, non-linear models
Reductionistic approach	Holistic approach

Box 2.1:
Mathematical Modelling

In the absence of complete knowledge and in order to guide their actions, people use simplified images of the surrounding world. One may call these images models of a, usually small, part of reality (Rotmans & de Vries, 1997). A model is not intended to replicate all components of a real system, but is designed to extract key features needed to answer specific questions (Aron & Silverman, 1994). This book confines itself to conceptual models and mathematical models. Conceptual models represent the system's boundaries, the essential entities of the system and their interrelationships. Mathematical models are conceptual models in which entities and interrelationships are formally represented by variables and relationships, often in the form of a set of differential or difference equations (Rotmans & de Vries, 1997).

Since complex mathematical models cannot usually be solved analytically, a computer simulation program is needed to solve the set of equations. Several general lines are followed during the mathematical modelling process (Janssen *et al.*, 1990):

- Problem formulation: Concerning the verbal definition of the problem.
- Problem analysis and structuring: In which the relevant parts (systems) of the practical situation are defined.
- Information sampling: Determination of the way the system components influence each other and collection of relevant data.
- Mathematical modelling: Representing the model in mathematical language.
- Model implementation: Translation of the mathematical symbols in a computer language.
- Program verification: Checking for possible difference between the mathematical model and the computer program.
- Model calibration: Selecting parameter and initial values of variables so that the model replicates the real world problem as closely as possible.
- Model analysis: Determining the model's properties, e.g. uncertainty analysis.
- Model validation: Checking of the conceptual (Are the assumptions and scientific theories indeed valid in the model?), practical (Does the model outcome reflect measured data?), and operational validity (Does the model give answers to the objectives of the study?).
- Model use: If the preceding steps are performed well, and the quality of the model seems good, the real application of the model may start.
- Results presentation and evaluation: Presenting the outcomes of the model in e.g. graphs and tables.

In practice, this process is an iterative one, and often two or more of the above mentioned steps are combined.

Furthermore, climate change and ozone depletion would not affect human health in isolation, but simultaneously and in conjunction with other ecological and demographic changes. Therefore, the net impact of global atmospheric changes would depend on various interactive phenomena: multiplicative exposure effects, feedback pathways, and differences in the vulnerability of (local) populations.

It seems that three major polarities characterise this new research domain in comparison to conventional epidemiology: (i) spatial scale, i.e. regional/global versus local impacts; (ii) temporal scale, i.e. future versus present health risks; and (iii) level of complexity, i.e. complex eco-epidemiological processes versus straightforward cause-effect relationships. Table 2.1 summarises the main differences between conventional epidemiology and an eco-epidemiological framework, as discussed above.

Current mainstream epidemiological research methods appear not well adapted to the analysis of disease causation which involves complex systems influenced by human interventions or more 'simple' processes which will take place within the (distant) future. Nor are the empirical sciences able to deal with uncertainties arising from such complex systems (Risk Assessment Forum, 1992; Levins, 1995; McMichael & Martens, 1995; Patz & Balbus, 1996). Thus, new approaches to estimating the health impacts of global climate change and ozone depletion are necessary. In this book, this perspective is labelled with the term 'eco-epidemiology'.

Eco-epidemiology has been used by others, for example to refer to studying the health impacts of chemical pollution of local/regional environments (e.g. the work on the Great Lakes pollution, with its impacts on fish and bird life, and indirectly, on humans (Carson, 1962)), or to refer to the need to give better consideration to human disease origins in a social ('human ecology') context (Susser & Susser, 1996a, b). In this study the term is used to describe an approach that should be capable of taking account of *ecological complexities*, and – at least as importantly – one that is able to estimate *future* health risks, with maximum reference to existing epidemiological knowledge of disease causation. Within this eco-epidemiological approach most quantitative estimates of future health impacts of global atmospheric changes will come from integrated mathematical computer modelling (McMichael, 1993; McMichael & Martens, 1995; Patz & Balbus, 1996) (Box 2.1).

ECO-EPIDEMIOLOGICAL MODELLING

MIASMA

MIASMA[1] is an acronym devised to refer to the models described in this book: the vector-borne disease model, the thermal stress model, and the skin cancer model. This modelling framework is designed to describe the major cause and effect relationships between atmospheric changes and human population health. The models (which are discussed in detail in terms of their aggregation level, temporal and spatial scale, etc., in the following chapters) are driven by scenarios of population figures and atmospheric changes, superimposed on baseline data regarding disease incidence, climatic conditions, and ozone-layer thickness (Martens et al., 1995c; Martens, 1996b). Global atmospheric changes directly influence the exposure to health risks, via changes in ambient temperature and received UV-B radiation, as well as indirectly, in influencing the dynamics and distribution of vector-borne diseases. Changes in the pattern of health risks demarcate the changes in the levels of incidence of the diseases influenced by the determinants. The mortality rates associated with cardiovascular diseases are directly influenced by thermal stress, mainly in urban areas. Figure 2.1 illustrates an outline of this modelling framework in conceptual terms.

The modelling approach is orientated towards a vertical integration of global atmospheric disturbances and their respective health effects. The models try to cover as much as possible of the cause-effect relationship with respect to global atmospheric changes and human health. In the vector-borne diseases model (see Chapter 3), the dynamics of malaria, schistosomiasis, and dengue, are simulated in relation to climate changes. Relationships between temperature, precipitation, and vector characteristics are based on a variety of field and laboratory data.

Changes in transmission dynamics of malaria and schistosomiasis are modelled using the basic infectious disease models described in Anderson & May (1991); for dengue a well validated, dynamic life-history model of dengue transmission (Focks et al., 1993a, b) is used. Recognising the need for continuing cross-validation of large-scale and small-scale studies (Root & Schneider, 1995),

---

[1] Miasma is a word derived from the Greek meaning: to pollute. An antique theory asserted that gases which rose from the soil, especially when high temperatures caused decomposition and decay to take place, contained 'miasma', a substance causing diseases and epidemics. Although the miasmatic theory was abandoned long ago, it is an irony of history that state-of-the-art environmental health science is now striving to quantify the relationship between atmospheric conditions and the incidence of diseases. In this study, 'miasma' is just used as a nice acronym with no relation to this old epidemiological paradigm.

simulations have been performed of the transmission potential of malaria in Zimbabwe and dengue in five cities (Bangkok, San Juan, Mexico City, Athens, and Philadelphia). The historical data available for these locations are used for validation, i.e. testing the performance of the model. Furthermore, the effects of using insecticides and antimalarial drugs on the incidence of malaria (and the development of resistance to them) are examined within a hypothetical model setting (Chapter 4).

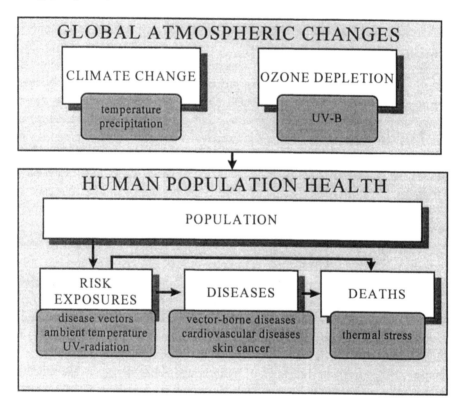

Figure 2.1: Conceptual Representation of MIASMA

To represent a wide range of climatic conditions and levels of socio-economic developments, effects of thermal stress on cardiovascular, respiratory and total mortality have been simulated for cities throughout the world (see Chapter 5). The association between winter and summer temperatures and mortality rates has been estimated by means of a meta-analysis, aggregating the results of several epidemiological studies on the subject. Projections of future risks are then simulated by simple extrapolation of this calculated relationship. Effects of acclimatisation to increasing temperatures, physiological as well as technological, are simulated. Allowing for the several delay mechanisms in the process of ozone depletion and skin cancer incidence, changes in skin cancer rates are simulated for The Netherlands and Australia, using a model first described in Slaper *et al.* (1992) (see Chapter 6). Model simulations are performed using various assumptions of ozone depletion, exposure habits and the sensitivity of the populations at risk.

Integrated Assessment Models

Although mathematical modelling is often used by epidemiologists – to gain insights into the observed dynamics of infectious disease epidemics, for example, or to estimate future time trends in diseases – the complex task of estimating future trends and outcomes in relation to global atmospheric change and human health requires the use of integrated, systems-based mathematical models (Rotmans *et al.*, 1994; McMichael & Martens, 1995). An integrated approach is the most comprehensive treatment of the interactions between atmospheric changes and society (Carter *et al.*, 1994). In general, integrated assessment models try to describe quantitatively as much as possible of the cause-effect relationship of a phenomenon (vertical integration), and the cross-linkages and interactions between different issues (horizontal integration), including feedbacks and adaptations. In practice, since the knowledge base is insufficient to permit us to conduct a full integrated assessment, only partial integration is possible. In many cases, partial integrated assessment models strongly resemble a straightforward 'cause and effect' approach (in this approach it is assumed that other factors on the exposure unit, such as non-climate factors, are held constant or an 'interaction' approach (the interaction approach recognises that climate, for example, is only one of a set of factors that influence or are influenced by the exposure unit) (Carter *et al.*, 1994), and the distinction is often difficult to make.

Although various modelling paradigms can be distinguished in the integrated assessment community (for a discussion of these paradigms see Meadows & Robison (1985), Janssen (1996), and Box 2.2.), in this book the following modelling approaches are described: the integrated systems approach in the case of the vector-borne diseases and skin cancer models, and the complex adaptive systems approach for the simulation of resistance development of the malaria mosquito and parasite; estimates of the effects of thermal stress-related mortality changes are generated by simple extrapolation.

Systems Approach

A systems approach is concerned with modelling real-world systems and studying their dynamics. Integrated modelling, which builds on systems-oriented analyses, concentrates on the interactions and feedback mechanisms between different subsystems of the cause-effect chain (rather than focusing on each subsystem in isolation) (Dzidonu & Foster, 1993). Feedback processes can amplify or dampen important aspects of the system. For example, an important determinant of the number of people infected by malaria is the level of (temporary) immunity within the target population. Hence, in highly endemic regions with a high prevalence of immunity, the impact of a climate-related increase in the malaria transmission potential of the mosquito population will be lower (and will soon be counteracted by the further boost in immunity (see also Chapter 3)) than the impact in populations with initially low levels of immunity.

Given the complexity of the phenomena affecting health that are at issue here, and our relative ignorance about the basic processes and interactions that determine their dynamics, the integrated systems approach can help to foster an understanding of the causal relationships that are responsible for changes in the structure and dynamics of the system. Therefore, the systems approach seems to be an appropriate method to capture the complexity of the interrelationships between global environmental changes and human health (the specific applications of this approach are discussed in Chapters 3 and 6).

Complex Adaptive Systems

Representing the 'world' in a pseudo-mechanical way, i.e. ignoring the system's ability to respond to changes, will often be inadequate. Von Bertalanffy (1968) noted that, while the systems approach can be used for describing the maintenance of a system, it cannot explain change, diversity and evolution. Because living organisms, such as human beings, animals, ecosystems, and societies, can respond, react, learn, adapt and influence each other, the taxonomy of the systems under consideration is likely to change during the (often) long period being simulated. Therefore, scanning the future of systems for the next century, without considering the ability of the systems to adapt to changes, may generate a misleading picture of the impacts of these changes.

The use of *complex adaptive systems*, which simulate such evolutionary processes as learning and adaptation and include continuous changing of the underlying systems, may be essential in the assessment of global change impacts (McMichael & Martens, 1995; Janssen, 1996). In recent decades, new computer-based modelling tools have been developed which enable these complex adaptive systems to be studied, among which are: genetic algorithms, cellular automata, artificial life forms and non-linear dynamic systems. Such evolutionary modelling techniques have been applied in various disciplines which study, for example, economies, ecologies and immune and nervous systems (e.g. Langton, 1989; Arthur, 1990; Kauffman, 1991; Holland, 1992; Engelen *et al.*, 1995). Although this new modelling paradigm is likely to derive new insights from various complex systems, it still has to prove itself (Horgan, 1995).

Insects, which develop resistance to a variety of pesticides, and the development of resistance to the drugs used in treatment in various disease-causing parasites, are examples of systems adapting to a changing environment. In Chapter 4, a first sketch is made of the development of an evolutionary modelling approach to the adaptation of the malaria mosquitoes and parasites to climate change, insecticides, and antimalarial drugs.

Box 2.2:
The Evolution of Modelling
(Source: Janssen (1996))

Newton's publication of his *Principia* in 1687 is generally seen as marking the birth of classical science, and this mechanistic, reductionistic and equilibrium-based explanation of the world was to prove successful in stimulating physical science. During the so-called Machine Age, research was dominated by selecting what was to be understood, studying the working of the parts, and assembling the understanding of the parts into an understanding of the whole. This came naturally to men whose view of the universe inspired them to create machines to do their (physical) work, and a product of such efforts was the Industrial Revolution itself. Moreover, the success of the mechanistic paradigm led to the application of the mathematical tools of physical science to life sciences such as economics, social science and biology. Although the success of classical science was to be threatened by developments in thermodynamics in the 19th century and challenged in the first half of this century by new theories in physical science (quantum mechanics and relativity theory) the reductionistic and deterministic view nevertheless remained the core of science.

By the second quarter of this century, formulation of a new world view was brought about in part by the growing preoccupation with systems, with their growing complexity and the increasing difficulty of managing them effectively. This focus on systems led to the realisation that they constitute wholes which lose their essential properties when taken apart. In metaphoric terms, the systems approach follows a holistic view instead of the reductionistic view of classical (Newtonian) science.

Nevertheless, the dominant world view was still ruled by deductive logic and mathematics: the rationalistic world view. A reversible world in which initial conditions determine a reversible trajectory is postulated so that once the initial state of the system has been determined, deductive logic can chart the past and predict the future trajectories. Whatever has changed can be exactly undone by another change, and using exact prediction, the behaviour of nature can be precisely controlled. However, Von Bertalanffy (1968) noted that while dynamic systems can be used for describing the maintenance of a system, they cannot explain change, diversity and evolution, and, since as early as 1948, Weaver (1948) has been stressing the need for an approach which deals with a *sizeable number of factors which are interrelated into an organic whole*, pointing to biological, medical, economic and social issues.

Box 2.2 Continued

There has thus been a growing realisation that if long-term processes in social and biological systems are to be studied effectively, the previous approaches are of limited use since societies, humans and other organisms live in continuous and changing interactions with their environment, leading to structural changes. Furthermore, since each biological or social agent has specific characteristics, the survival of the fittest means that individuals are not equally successful. In metaphoric terms: we shift from the system as a machine to the system as an organism.

Some major differences between the 'Newtonian' approach, the systems approach and the complex adaptive systems approach, are as follows: The Newtonian approach offers us universal laws, and the systems approach a holistic general cause-effect diagram of the system, while complex adaptive systems deal with the diversity of individual characteristics of the agents. If the Newtonian approach could describe how systems move to an equilibrium, and complex systems the pathways between multi-equilibria, complex adaptive systems, in contrast, are seen as evolving over time, thereby adapting to the continuous changes within the system.

The mechanistic perspective of Newtonian science and the systems approach allow us to construct models in order to control the system optimally and to predict future developments. According to the complex adaptive systems approach, however, changes in structure and behaviour can neither be predicted nor fully controlled.

The information exchange between the various elements in the system differs between the approaches. The Newtonian approach assumes perfect knowledge leading to an equilibrium state of the system. In the (complex) systems approach, the information about the state of various parts of the system is modelled as feedback mechanisms. In the complex adaptive systems approach, the agents have preferences and expectations leading to the anticipation of their behaviour. Differences in expectations and realised behaviour of the system lead to adaptive behaviour or physical adaptation of the agents.

The complex adaptive systems approach implies a novel way of dealing with models, a problem which is currently underestimated. An approach using models will be required which accommodates the notion that the future is not only uncertain, but also inherently unpredictable. This may yield insights, but not the precise answers claimed by Newtonian science.

Limitations and Advantages of Integrated Assessment Models

Integrated assessment models, based either on a systems approach or on a complex adaptive systems approach, may not pretend to offer a comprehensive picture of all relevant processes of complex realities. In view of the accumulation of uncertainties, which is inherent in integrated assessment modelling, the interpretative and instructive value of the models presented in the following chapters is far more important that their predictive potency, which is limited by the incomplete science on which they are constructed. According to Rotmans *et al.*

(1996), the limitations and drawbacks of integrated assessment models may include the high level of aggregation, the absence of stochastic behaviour, the limited possibilities for validation, and inadequacy of knowledge and methodology. Furthermore, it is important to note that models using either one of the approaches (i.e. the systems-based dynamic or the complex adaptive systems approach) should be as simple as is both realistic and possible, especially since uncertainties in such models cannot be reduced by increasing their complexity. Furthermore, it is important that the assumptions and limitations of such models are made explicit.

The major advantages of integrated assessment models (Rotmans *et al.*, 1996) are: (i) systems are included in interactions and feedback mechanisms; (ii) the simplified nature of the modules in integrated models permits rapid prototyping of new concepts and exploration of their implications; (iii) uncertainties, crucial lacunae in current scientific knowledge, and weaknesses in discipline-oriented expert models can be identified and revealed; (iv) the accumulation of uncertainties can be analysed and interpreted; and (v) integrated models are outstanding means of communication between scientists and exponents of many disciplines, and between scientists and decision-makers. Furthermore, integrated assessment models may serve as a repository of what is known about the elements of a system and their relationships, and can augment extrapolation from historical data. In the following sections, some of the issues described above are discussed in greater detail.

Aggregation Level
One of the critical issues in integrated assessment modelling is that of aggregation versus disaggregation (Rotmans *et al.*, 1996). The level of aggregation within a modelling framework refers to the formulation of the dynamics in the model in terms of complexity, and the level of detail, which is often closely related to the spatial and temporal resolution chosen within the framework. The problem with integrated assessment models is that they often consist of a variety of submodels, which have different aggregation levels; in other words, they differ in complexity and spatial and temporal resolution, etc.

The assessment of health vulnerability due to stratospheric ozone depletion and climate change may be done on a variety of geographical scales, varying from a village to an entire country, region, or the world as a whole. Furthermore, the response time of human systems to atmospheric changes also differs between diseases and locations. For example, climatic changes simulated with General Circulation Models (GCMs) have a relatively coarse spatial resolution – grid cells of a few degrees – but run at a fine temporal resolution; ecological models mostly require data of a fine spatial resolution, but their time resolutions may vary from one day to a season or a year. In general, because of the multiple ecosystem levels which must be altered before human health is affected, there will be a time lag between the change of the environmental stressor and the health impacts (with the exception of thermal-related mortality, caused by more direct disease processes) (Patz & Balbus, 1996).

Calibration and Validation

As uncertainties as to the exact values of the parameters or the initial values of the time-dependent variable always remain, the calibration process is meant to adapt these values in such a way that the mathematical model adequately reflects the real-world problem. Thus calibration amounts to a comparison of model output and measured data and, as such, the calibration step is close to validation. Validation is defined here as the procedure for testing the adequacy of a given mathematical model (Rotmans et al., 1994). With respect to the calibration and validation of dynamic models, however, there are two major problems. First, complete calibration and validation of simulation models is impossible, because the underlying systems are never closed (Oreskes et al., 1994). Second, calibration and validation are often not possible because the requisite data and scientific knowledge are not available. There is a close interrelationship between the identification of data needs and the selection of methods of analysis. One of the problems often encountered in applying process-based models in less developed countries is that the models, often adequately validated in the data-rich developed world, are found to be ill suited to or poorly calibrated for use in the less developed countries. A paucity of data for validation generally means that data-demanding models can often not be used in such circumstances, and reliance has to be placed on less data-demanding models (Carter et al., 1994). Unavailability of data will necessitate a reliance on simplified assumptions to generate an initial framework for analysis; this framework can be used to focus interdisciplinary communication on assessing health risks and identify priorities for future research. Although the use of such assumptions and simplifications will potentially decrease the quantitative accuracy of the assessment, it should still allow for adequate prioritisation and estimation of relative risk (Patz & Balbus, 1996).

Validation can be divided into different types (Rodin, 1990; Rykiel, 1996). Data (or pragmatic) validation requires concordance of the model's projections with observational data sets; this concordance can often be assessed by 'testing' the model on historical data sets. However, as mentioned above, the relative inaccuracy and imprecision of eco-epidemiological data places limits on the model's testability. Conceptual validation requires that the hypotheses and the theoretical structures of the model reasonably describe the perceived real world. This implies that the model structure, relations, parameters and dynamic behaviour reflect the prevailing theoretical insights and the key facts relating to that part of reality that the model is supposed to represent. However, this does not necessarily mean that a conceptually valid model will make accurate projections. Finally, operational validation requires that the model fulfils the objectives of the study and permits conclusions to be drawn on the problem.

Uncertainties

Despite the many advantages of the integrated systems approach, one of its disadvantages is the sequence of uncertainties introduced by the linkage of separate modules: the uncertainty range widens as one moves to more remote links in the cause-effect chain. This cumulation of uncertainties is illustrated in the section on the assessment of the impacts of climate change on malaria prevalence (Chapter 3) and in the section on the impacts of increased UV-B radiation due to ozone depletion on skin cancer incidence (Chapter 6). On the other hand, integrated assessment models enable comparison of the relative importance of these uncertainties (McMichael & Martens, 1995).

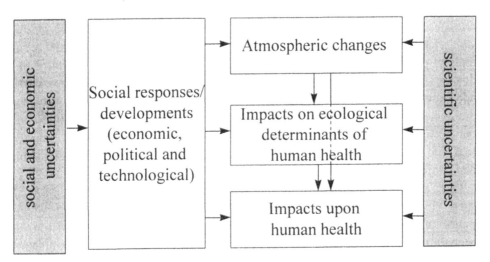

Figure 2.2: Layers of Uncertainty Underlying the Health Impact Assessment of Global Atmospheric Changes

The types of uncertainties, in the time-scale and magnitude of the incipient or anticipated disturbances of natural systems, and, therefore, in their impacts on population health, warrant further consideration. First, the projection of health impacts is contingent on a multi-layered infrastructure of uncertainties from other disciplines: climatology, atmospheric chemistry, agricultural science, ecology, social sciences, economics, and so on (Figure 2.2). From those disciplines come projections about atmospheric changes (temperature increase, ozone depletion, etc.). Further, we have little *directly* relevant empirical evidence about the impacts of many of these projected environmental changes on human health. The impact of such a change in background climate on human health can therefore often not be directly estimated by extrapolation from prior empirical observations of the health consequences of short-term temperature fluctuations around a stable mean temperature.

Many classifications of uncertainties can be made (e.g. Funtowicz & Ravetz, 1989; Morgan & Henrion, 1990). Here, the various types and sources of uncertainty are aggregated into two categories (Rotmans *et al.*, 1994). (i) Scientific uncertainties arise from the degree of unpredictability of global atmospheric change processes and their impact upon human health, and may be narrowed as a result of further scientific research or more detailed/appropriate modelling; (ii) social and economic uncertainties arise from the inherent unpredictability of future geopolitical, socio-economic, demographic and technological evolution. Scientific uncertainties include, for example, incomplete knowledge about the dose-response relationships between UV radiation and skin cancer incidence (Chapter 6), or the lack of knowledge about the shape of the function between the 'fitness' of a malaria parasite/mosquito and the dose of drugs/insecticides used to control the disease (Chapter 4). Examples of social and economic uncertainties are the cultural adjustments in time which may have an impact on the relationship between thermal stress and mortality rates, such as improvements in housing conditions and better clothing (Chapter 5). Another example is the fact that changes in lifestyle (e.g. sun exposure) strongly influence the dosage of UV-B received by the skin, and consequently skin cancer rates (Chapter 6).

There are several approaches to the analysis and presentation of uncertainties: (i) specifying a set of future scenarios; (ii) the parametric analysis, in which parameters' values are changed over a range of possible values; (iii) the probabilistic method, using a (subjective) probability distribution for empirical quantities. A simple way of presenting uncertainties is by specifying a set of future scenarios, where the scenarios selected are expected to span a range of plausible, representative futures. Another, more systematic method of uncertainty study is sensitivity analysis, which examines the sensitivity of outputs to variations in key inputs. The problem here is that it requires some prior judgement to select key inputs, and it teaches us more about the sensitivities than about the nature of the uncertainties. A more comprehensive approach to uncertainty is the probabilistic method, whereby large numbers of inputs are specified as probability distribution functions, and a number of repeated model runs are done to determine the uncertainties surrounding the output(s). The major difficulty with this method is that it requires specific knowledge about the nature of the distribution functions and the number of runs required.

DISCUSSION

A major task in the scientific investigation of public health issues is to provide policy-makers and their public constituency with a clearer description of the anticipated future health impacts of global atmospheric changes. The complexity of the mediating processes of health impacts of global climate change and atmospheric ozone depletion requires a systems-based, eco-epidemiological modelling approach, the prior information gleaned from conventional epidemiological research being essential in the modelling of future, larger-scale health impacts.

However, with respect to the models described in the following chapters, the first point to make is that aggregating data about the 'natural' world (including human populations) necessarily involves simplifications, summarises, and averaging. This is a truism for conventional epidemiology, and it is no surprise that eco-epidemiologically based modelling involves similar limitations.

A further difficulty of an eco-epidemiological modelling approach is that vulnerability to global atmospheric changes will vary greatly between different segments of the world's population. Poorly resourced populations, such as those of Bangladesh and sub-Saharan Africa, will be more vulnerable to adverse (climatic) events than will rich nations. Globally aggregated models average over all populations, but specific projections will often be required for more localised populations. Within MIASMA the modelling of future health impacts of climate change has been done both at a highly aggregated global (or broad regional) level (malaria, schistosomiasis and dengue transmission (Chapter 3)), as well as on a more local (e.g. city/country) level (impact assessment of malaria in Zimbabwe and dengue in selected cities (Chapter 3), thermal stress in selected cities (Chapter 5) and effects of increased UV radiation due to ozone depletion in The Netherlands and Australia (Chapter 6)).

A final point, as indicated in the previous chapter (and discussed further in the following chapters), is that even without climate change and ozone depletion the complexity of influences of various factors upon health defies a ready quantitative analysis of net effects. For example, the distribution of malaria is strongly influenced by the use of pesticides, the availability of vaccination and the emergence of drug resistance in the *Plasmodium* parasite. Hence, the extent of any increase in malarial risk due to climate changes will be superimposed on the change in malaria transmission associated with socio-economic development, population growth and the effectiveness of control measures. Although an attempt to analyse simultaneously effects of climate change and the development of drug and insecticide resistance is discussed in Chapter 4, this model is at present theoretical and has not been implemented for a 'field situation'.

As the *full* complexity of assessing the health impacts of climate change and ozone depletion cannot be satisfactorily reduced to mathematical modelling, what is the role of such modelling? Despite the difficulties and limitations of the modelling process, the models discussed previously first of all draw attention to the *fact* of there being a foreseeable health impact of these global environmental changes. Second, they indicate the relative importance of climate change and ozone depletion as an influence upon these outcomes. This should enhance public discussion, education and policy-making. However, even more important is the role of eco-epidemiological modelling in the systematic linkage of multiple cause and effect relationships based on available knowledge and reasoned guesses. This should increase our understanding of climate- and UV-related health impacts, and identify key gaps in data and knowledge needed to improve the analysis of these effects.

Chapter 3

CLIMATE CHANGE AND VECTOR-BORNE DISEASES

INTRODUCTION

The occurrence of vector-borne diseases extends from the tropics and subtropics to the temperate climate zones. With a few exceptions, vector-borne diseases do not occur in the colder climatic regions of the world. The incidence of vector-borne diseases is determined by various factors, such as human behaviour, vector characteristics, and the presence of the relevant parasite. Any factor influencing these determinants will influence the likelihood of disease transmission.

Figure 3.1 gives a survey of the impact of climate change on vector-borne disease transmission. Direct effects of the anticipated changes in global and regional temperature, precipitation, humidity and wind patterns resulting from anthropogenic climate change are factors that have an impact on the vectors' reproduction, development rate and longevity. These factors would thus be associated with changes in vector density. In general, the rate of development of a parasite accelerates as the temperature rises.

Indirect effects of climate change would include changes in vegetation and agricultural practices, which would mainly be caused by temperature changes and trends in rainfall patterns. These changes either promote or inhibit disease transmission by their association with increased or decreased vector density. Irrigated land (such as paddy fields) provides a suitable breeding ground for a number of vectors. Vector-borne diseases which are greatly affected by changes in irrigation practices and in the distribution of irrigated areas include malaria and schistosomiasis. In areas in which extensive use is made of pesticides, resistance among vectors to insecticides can occur, with major consequences for disease transmission. A further indirect effect of climate change would be associated with the rise in sea level and the resultant coastal flooding. The proliferation of lagoons containing brackish water influences the availability of habitat and either encourages or discourages vector species, depending on whether they can tolerate brackish water. Generally speaking, drought and desertification, including the migration or extension of global desert belts, could be expected to decrease vector-borne disease transmission. After all, vector breeding generally relies on an aquatic environment and drought conditions severely curtail the vector's longevity. Thus climate change may lead to a change in the location of habitats capable of supporting vectors.

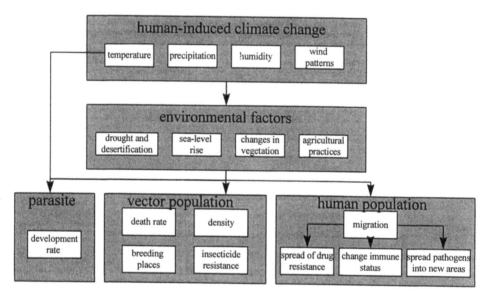

Figure 3.1: Schematic Diagram of the Major Climate Change Implications for Vector-Borne Diseases

The influence which climate change is likely to exert on human populations may also play an important role in the dynamics of disease transmission. For example, the large-scale migration of populations from areas in which vector-borne diseases are endemic into receptive areas (areas in which vector numbers and climate conditions are conducive for transmission) because of rural impoverishment, which is influenced by the dynamics of climate change (including the effects of sea-level rise on low-lying coastal areas), would prove significant.

Three of the world's most prevalent vector-borne diseases are malaria, schistosomiasis, and dengue, and there are few infectious diseases which have as great an impact on the social and economic development of human societies (Table 3.1). At present some 2400 million, 1800 million and 600 million people are regarded as being at risk of contracting malaria, dengue and schistosomiasis, respectively. Between 300 million and 500 million people are actually infected with the malaria parasite and the current prevalence of schistosomiasis is about 200 million. Between 10 million and 30 million dengue cases occur yearly throughout the world, making dengue by far the most important and the most geographically widespread of arboviral diseases (Halstead, 1993; Gubler & Clark, 1995).

Table 3.1: Global Status of the Major Vector-Borne Diseases in the World (Source: McMichael, (1996))

Disease	Vector	Population at risk (millions)	Prevalence (millions)	Present distribution	Likelihood of altered distribution with climatic changes
Malaria	Mosquito	2400	300–500	(Sub)tropics	+++
Schistosomiasis (bilharzia)	Snail	600	2007	(Sub)tropics	++
Dengue (breakbone fever)	Mosquito	1800	10–30 /year	All tropical countries	++
Lymphatic filariasis	Mosquito	1094	117	(Sub)tropics	+
African trypanosomiasis (sleeping sickness)	Tsetse fly	55	0.25–0.3 cases/year	Tropical Africa	+
Dracunculiasis (Guinea worm)	Crustacean	100	0.1 cases/year	South Asia/Arabian peninsula/central west Africa	?
Leishmaniasis	Phlebotomine sand fly	350	12	Asia/southern Europe/Africa/ Americas	+
Onchoceriasis (river blindness)	Black fly	123	17.5	Africa/Latin America	++
American trypanosomiasis (Chagas' disease)	Triatomid bug	100	18	Central and South America	+
Yellow fever	Mosquito	450	<0.005 cases/year	Tropical south America and Africa	+

?, Unknown; +, likely; ++, very likely; +++, highly likely.

While the sensitivity of malaria, schistosomiasis, and dengue to temperature and precipitation is relatively well documented, only a few studies have examined the potential influences of global anthropogenic climate changes on the distribution of these vector-borne diseases (e.g. Martens *et al.*, 1994, 1995a, b, 1997; Matsuoka & Kai, 1994; Martin & Lefebvre, 1995; Jetten *et al.*, 1996; Martens, 1996a; Patz *et al.*, 1998). The analyses described in this chapter do not claim that climate factors are the most important determinant of the vector-borne diseases considered. However, this chapter illustrates and quantifies the influence that the direct effects of climate change, as well as the indirect effects of changes in moisture (and the consequent changes in vegetation patterns), may contribute to altering the occurrence of malaria, schistosomiasis and dengue on a global scale. The results, rather than being predictive, should be interpreted as an indication of the sensitivity of the vector-borne diseases to climatic changes, temperature in particular. Future risk assessments of climate change will ultimately need to integrate (global) climate scenario-based analysis, such as the one presented here, with local socio-economic and environmental factors to guide comprehensive and sustainable preventive health strategies.

THE VECTORS AND THEIR DISEASES

Malaria is caused by species of parasite which belong to the genus *Plasmodium*, and the vector responsible for malaria transmission is the mosquito of the genus *Anopheles*. *Anopheles* mosquitoes belong to a very large genus which includes hundreds of species throughout the world, although only 60 of these are actual or potential malarial vectors. Some species prefer to take their blood meals from animals (zoophilic) and thus transmit malaria to humans very rarely or do not live long enough to allow the parasites to multiply and to develop inside them, while in some species the parasites seem to be incapable of development (Gillies, 1988). Although *Anopheles* mosquitoes occur most frequently in tropical or subtropical regions they are found in temperate climates and even in the polar regions during summer. In countries in which malaria has been eradicated, mosquito vectors capable of transmitting malaria nevertheless still exist. There are four species of the malaria parasite: *P. vivax*, having the most extensive geographic range; *P. falciparum*, the most common species in tropical areas and the most dangerous clinically; and *P. ovale* and *P. malariae,* which are less prevalent.

The life cycle of the malaria parasite involves transmission both from mosquito to man and from man to mosquito, effected by the bite of a female mosquito. The parasite multiplies within the mosquito by means of sexual reproduction, and following an incubation period of several days (depending on the temperature and the species of parasite), malarial parasites can be found in the salivary glands of the insect. When an infected mosquito bites a human host, saliva is also injected and parasites are thus transferred to (hitherto uninfected) people. Asexual multiplication takes place in the human host. Having received an infective bite, there is an incubation period in the patient which varies between 10 and 40 days, depending on the species of parasite. Towards the end of the incubation period the infected person may suffer from headaches, pains in the arms and legs, backache, nausea and vomiting. The incubation period culminates in a severe attack which is caused by the destruction of infected blood cells and the release of toxins into the bloodstream. Infections involving *P. falciparum* are often associated with fatal complications (e.g. anaemia and cerebral malaria).

The transmission of dengue is comparable to the transmission of malaria: here mosquitoes of the species *Aedes aegypti* are the principal vectors. The dengue viruses belong to the family of *Flaviviridae* and the genus *Flavivirus*. There are four dengue serotypes (designated DEN-1, -2, -3, and -4). Although immunity may be acquired after infection has taken place, there is limited cross-protective immunity between the different dengue serotypes. Dengue infection causes a spectrum of illness in humans ranging from clinically unapparent to severe and fatal haemorrhagic disease; the acute illness lasts for about 3–7 days. The case fatality rate of the more severe form, dengue haemorrhagic fever/dengue shock syndrome (DHF/DSS), approximates 40–50 per cent if untreated (Beneson, 1990).

The transmission cycle of schistosomiasis is more complex than that of malaria and dengue. Three species of schistosome account for most human schistosomiasis: *Schistosoma mansoni*, *S. japonicum* and *S. haematobium*. The different species of schistosome that infect humans have similar life cycles. The eggs hatch in water as free-swimming larvae called miracidia. A miracidium must penetrate an appropriate snail within a certain time span (approximately 32 hours). Each species of schistosome can infect only a single species of snail. Therefore, the possible transmission of each form of schistosomiasis depends on the presence of a suitable host. The snail genera responsible for schistosomiasis transmission are *Biomphalaria* (*S. mansoni*), *Bulinus* (*S. haematobium*) and *Oncomelania* (*S. japonicum*) (Jordan & Webbe, 1982).

Asexual reproduction of the schistosome takes place in the snail. A single miracidium produces about 200–400 free-swimming larvae, called cercariae, that are shed from the snail and must penetrate the skin of a human host within approximately 2 days. The penetration of the skin takes about 2 minutes. After penetration, the schistosome migrates through the host's circulatory system to the liver. On reaching maturity and mating, the schistosome migrates to the veins in the intestines (*S. mansoni* and *S. japonicum*) or to the veins in the bladder (*S. haematobium*). From here the eggs are deposited through faeces or urine into the water (Weil & Kvale, 1985). The majority of the persons infected do not present symptoms of the disease. Symptoms do not result from the adult worm, but rather from the eggs that remain in the host tissue.

The health complications appear to vary according to the species and strain of parasite and the characteristics of the human population. Schistosomiasis usually causes blood and nutrient loss in either the stool or the urine, which may lead to anaemia and retarded physical growth. More serious complications include bladder or ureter calcification in urinary schistosomiasis and an enlarged liver and spleen in intestinal schistosomiasis. An association between schistosomiasis and bladder and colorectal cancer has also been found (WHO, 1993a).

Figure 3.2 depicts the transmission cycles of malaria, dengue and schistosomiasis and the interaction of temperature and precipitation as they affect the various stages in these cycles. The effects of temperature on disease transmission will be discussed in detail in the next section.

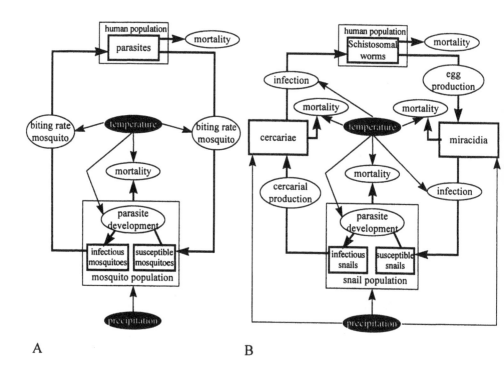

Figure 3.2: Diagram of Temperature and Precipitation Effects on Main Population and Rate Processes Involved in the Life Cycle of the Malaria and Dengue (A), and Schistosome (B) Parasite

EPIDEMIC POTENTIAL

The dynamics of the malaria, dengue and schistosomiasis vector populations are much more rapid than human population dynamics, so the vector dynamics can be considered in equilibrium with respect to changes in the human population. A unit of measurement which encapsulates many of the important processes in the transmission of infectious diseases is the basic reproduction rate (R_0). In the case of the malaria protozoan and dengue viral microparasite, R_0 could be more precisely defined as the average number of secondary infections produced when a single infected individual is introduced into a potential host population in which each member is susceptible. R_0 is closely related to the vectorial capacity (VC), a unit of measurement often used in malaria epidemiology: R_0 is the VC multiplied by the duration of the infectious period in humans. In the case of the transmission of macro-parasitical schistosomiasis worms, which involves two intermediate larval stages, R_0 is the average number of female offspring produced throughout the lifetime of the female parasite (Anderson & May, 1991). To put it in simple terms, if $R_0 > 1$ the disease will proliferate indefinitely; if $R_0 < 1$ the disease will die out (see Box 3.1).

The use of the basic reproduction rate depends on the assumption of homogeneous mixing, such as the assumption that vectors die at a constant rate, independent of age. Dye (1986, 1990) points out that at best, an estimate of the VC (or basic reproduction rate) will be useful as a "comparative index changing proportionally with the true vectorial capacity from site to site, from vector to vector, and within and between transmission seasons". Nevertheless, the basic reproduction rate provides a means by which to study the effect of climate change on vector-borne diseases' epidemiology.

Vector density is one of the parameters involved in the basic reproduction rate, which is strongly related to local environmental conditions. The change in the numbers of prevailing malaria, dengue and schistosomiasis vectors over time varies greatly between species, being determined by numerous biological and physical factors, such as the availability of species-specific breeding sites. It is impossible to arrive at a reliable estimate of the change in vector density over large areas as a result of changes in temperature, precipitation and humidity changes in aggregated models such as the ones presented in this chapter. Nevertheless, the formula for the basic reproduction rate allows calculation of the critical density threshold of vector populations necessary to maintain parasite transmission. The critical density (m_{c1} in mosquitoes per human) for malaria and dengue transmission can be expressed as (Dietz, 1988):

$$m_{c1} = c_1 \frac{-\ln(p)}{bca^2 \, p^n} \tag{3.1}$$

where p is the survival probability of the mosquito; a the frequency with which human blood meals are taken; n the incubation period of the parasite in the vector; b the efficiency with which an infective mosquito infects a susceptible human; and c the efficiency with which an infected human infects a susceptible mosquito. The term c_1 is a function, assumed to be constant, that incorporates the recovery rate in man; the duration of infection is more dependent on innate pathogen or host characteristics than on external climate or ecological factors.

In the case of schistosomiasis, which is transmitted indirectly between hosts in the sense that free-swimming stages are interposed, the idea of a critical density (in snails times humans) may be expressed as (Anderson & May, 1985):

$$m_{c2} = c_2 \frac{\mu_{mir} \, \mu_{cer} \, \mu_{ss}(\mu_{sls} + \sigma)}{\beta_{human} \, \beta_{snail} \, \sigma} \tag{3.2}$$

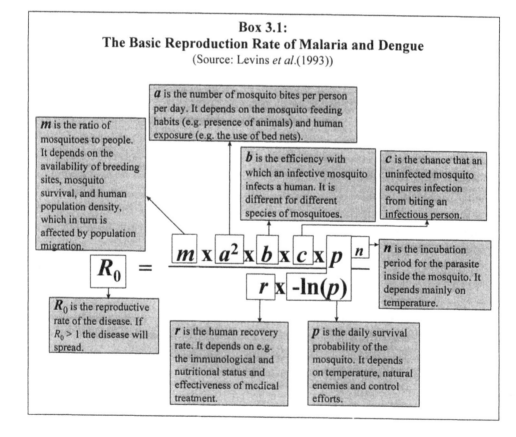

Box 3.1:
The Basic Reproduction Rate of Malaria and Dengue
(Source: Levins et al.(1993))

a is the number of mosquito bites per person per day. It depends on the mosquito feeding habits (e.g. presence of animals) and human exposure (e.g. the use of bed nets).

m is the ratio of mosquitoes to people. It depends on the availability of breeding sites, mosquito survival, and human population density, which in turn is affected by population migration.

b is the efficiency with which an infective mosquito infects a human. It is different for different species of mosquitoes.

c is the chance that an uninfected mosquito acquires infection from biting an infectious person.

$$R_0 = \frac{m \times a^2 \times b \times c \times p^n}{r \times -\ln(p)}$$

n is the incubation period for the parasite inside the mosquito. It depends mainly on temperature.

R_0 is the reproductive rate of the disease. If $R_0 > 1$ the disease will spread.

r is the human recovery rate. It depends on e.g. the immunological and nutritional status and effectiveness of medical treatment.

p is the daily survival probability of the mosquito. It depends on temperature, natural enemies and control efforts.

where μ_{mir}, μ_{sls}, μ_{ss} and μ_{cer} are the death rates of the miracidia, uninfected (susceptible and latent) snails, infected (shedding) snails, and cercariae, respectively. The rates of human and snail infection by cercariae and miracidia, respectively, are given by β_{human} and β_{snail}. The term $1/\sigma$ expresses the latent period of the parasite inside the snail. Like c_1 above, c_2 is a constant function representing factors which are assumed to be independent of temperature changes, which, in the case of schistosomiasis, include the death rate among the mature worms in man, the rate of egg production by female worms, the rate of cercarial production per snail, human sanitation habits and water contact rates, and the probability that a female worm is fertilised within the human host. This probability of fecundity approaches unity in most endemic areas (Anderson & May, 1991).

The epidemic potential (EP) of malaria, dengue, and schistosomiasis is defined as the reciprocal of the host density threshold. This EP is a key summary parameter which is used as a comparative index in estimating the effect on the risk of malaria, dengue and schistosomiasis represented by a change in ambient temperature and precipitation patterns. A high EP indicates that despite a smaller vector population, or, alternatively, a less potent vector population, a given degree of endemicity may be maintained in a given area.

CLIMATE EFFECTS

The vectors which transmit malaria, dengue, and schistosomiasis live in a variety of natural habitats in which conditions may vary widely, and their distribution and population dynamics are probably governed more by abiotic than by biotic factors (Southwood, 1977). Among the possible abiotic influences on the transmission cycles described above, the most important are temperature and rainfall. While some of the effects of climate changes on vector-borne diseases are highly local in character, others, particularly the effects of rising temperature, are more general (i.e. not place-specific) (Bradley, 1993).

The influence of climatic conditions on the malarial and dengue mosquito, schistosome snail, and their respective parasites has been addressed in numerous studies, in most of which attention is restricted to the effect of temperature on a single genus of parasite or vector. Although separate, detailed data on each species are usually absent, we assume that the relationships between climate parameters (in particular temperature) and specific parasite and vector characteristics hold for other species as well. In the case of schistosomiasis, the empirical relationships between temperature and transmission dynamics are consistently defined with respect to *water* temperature. Extreme temperatures are seldom a limiting factor for the snail vector (except when the habitat is drying out), as water may be several

Table 3.2: Main Temperature-Dependent Parameters in the Malaria, Dengue, and Schistosomiasis model

Parameter	Depending on		Default value in model	
Malaria				
n (latent period parasite)			T-dependent	
	D_m	(degree days parasite development)	105	°C day (*P.vivax*)
			111	°C day (*P.falciparum*)
	$T_{min,m}$	(minimum temp. parasite development)	14.5	°C (*P.vivax*)
	,		16	°C (*P.falciparum*)
a (man-biting rate)			T-dependent	
	HBI	(human blood index)	0.4	
	FI	(feeding interval)	T-dependent	
	- D_{bd}	(degree days blood digestion)	36.5	°C day
	- $T_{min,bd}$	(minimum temp. blood digestion)	9.9	°C
p (survival probability)			T-dependent (max. ~0.9/day at 20°C)	
b (human susceptibility)			1	
c (mosquito susceptibility)			1	
Dengue				
n (latent period parasite)			T-dependent	
		minimum temperature virus development	11.9	°C
		virus titer	5.4	MID_{50}
		(see also under 'a')		
a (man-biting rate)			T-dependent	
	r	(development rate)	T-dependent	
	- ρ (25°C)	(development rate at 25°C)	0.003359	/h (virus dev)
			0.00898	/h (blood dig)
	- $\Delta H_A^{\neq}$	(enthalpy of activation)	15,000	cal/mol(virus dev)
			15,725.23	cal/mol(blood dig)
	- $\Delta H_H^{\neq}$	(enthalpy of inactivation)	6.203e30	cal/mol(virus dev)
			1,756,481.07	cal/mol(blood dig)
	- $T_{1/2H}$	(50% inactivation temperature)	-2.176e30	K (virus dev)
			447.17	K (blood dig)
	- R	(universal gas constant)	1.987	cal/(mol deg)
		cumulative development first gonotrophic cycle	1.0	
		cumulative development subsequent cycles	0.58	
	HBI	(human blood index)	0.9	
		feeding frequency per gonotrophic cycle	T-dependent	
		- female's weight	T-dependent	
		- alternate host per feeding attempt	2.8	persons
p (survival probability)			T-dependent	(max.0.89/day between 6–40°C)
b (human susceptibility)			0.9	
c (mosquito susceptibility)			0.45	
		virus titer	5.4	MID_{50}
Schistosomiasis				
n (latent period parasite)			T-dependent	
	D_{st}	(degree days schistosome development)	268	°C day
	$T_{min,st}$	(min. temp. schistosome development)	14.2	°C
μ_{mir} (death rate miracidia)			T-dependent	
μ_{cer} (death rate cercariae)			T-dependent	
β_{snail} (infection rate of miracidia)			T-dependent	
β_{human} (infection rate of cercariae)			T-dependent	
μ_{sls} (death rate non-shedding snails)			T-dependent	
μ_{ss} (death rate shedding snails)			T-dependent	

degrees cooler than the air temperature on very hot days. However, this difference is far smaller when monthly mean temperatures are considered. Therefore, and because the thermal conditions of *shallow* waters usually reflect the ambient temperature of the air (Baptista *et al.*, 1989), our simulations use ambient air temperature as an approximation of the water temperature. However, aquatic snails (and the miracidia and cercariae), as well as mosquito vectors, can select micro-habitats where temperatures are more suitable. Table 3.2 summarises the main temperature-dependent parameters of malaria, schistosomiasis and dengue, used in the model simulations, which will be discussed in the next section. Parasite development in both the mosquito and the snail tolerates a more restricted range of temperature than the intermediate hosts. This stage is sensitive to climate conditions and can be considered as the 'weakest link' in the transmission chain as far as temperature is concerned.

Temperature Effects

Malaria

The probability that a successful transmission of sporozoites to a susceptible human will develop into a malaria parasitaemic (b) depends partly on immune processes in and genetic differences between humans. Furthermore, the ability of mosquitoes to transmit different species of parasites, as denoted by the parameter c in equation 3.1, shows a wide variability (Nedelman, 1985; Molineaux, 1988). Because there is no information available about the dependency of both b and c on meteorological conditions, these parameters have been set to 1 (Jetten *et al.*, 1996).

The incubation period of the parasite in the malarial mosquito (the extrinsic incubation period (EIP)) must have elapsed before the infected vector can transmit the parasite. The parasites develop in the vector only within a certain temperature range, where the minimum temperature for parasite development lies between 14.5°C and 15°C in the case of *P. vivax* and between 16°C and 19°C for *P. falciparum*, while the proportion of parasites surviving decreases rapidly at temperatures over 32°C–34°C (Horsfall, 1955; Macdonald, 1957; Detinova *et al.*, 1962). The relation between ambient temperature and latent period is calculated using a temperature sum as described by Macdonald (1957):

$$n = \frac{D_m}{T - T_{min,m}} \tag{3.3}$$

where n is the incubation period of the parasite inside the vector (in days), D_m the number of degree days required for the development of the parasite (=105°C days and 111°C days for *P. vivax* and *P. falciparum*, respectively (Detinova *et al.*,

1962)), T the actual average temperature (between $T_{min,m}$ and a maximum temperature of about 40°C) and $T_{min,m}$ the minimum temperature required for parasite development (14.5°C and 16°C for *P. vivax* and *P. falciparum*, respectively) (see Figure 3.3A).

The number of blood meals a mosquito takes from human beings (a (1/day)) is the product of the frequency with which the vector takes a blood meal (1/feeding interval (FI, days)) and the proportion of these blood meals that are taken from humans (the human blood index (HBI)):

$$a = \frac{\text{HBI}}{\text{FI}} \qquad\qquad (3.4)$$

The HBI is the estimated proportion of the blood meals taken by a mosquito population which are obtained from humans (Garrett-Jones, 1964; Garrett-Jones *et al.*, 1980) and provides an indication as to whether a mosquito species is anthropophilic in its feeding behaviour (a high HBI indicates a preference for biting man) or zoophilic (a low HBI: species feeds mainly on animals). The index may vary between species, between locations, and over time. Furthermore, the availability of alternative hosts and environmental conditions influence the index. An accurate estimate of the HBI for a species is thus difficult to give. The HBI is set to 0.4, a value frequently found in malaria-endemic regions for a wide range of anopheline species (Garrett-Jones *et al.*, 1980). The frequency of feeding depends mainly on the rapidity with which a blood meal is digested (Detinova *et al.*, 1962; Service, 1980), which increases as temperature rises, and can be calculated by means of a temperature sum (Figure 3.3B) (Detinova *et al.*, 1962):

$$\text{FI} = \frac{D_{bd}}{T - T_{min,bd}} \qquad\qquad (3.5)$$

where D_{bd} is the number of degree-days required for the digestion of a portion of ingested blood (36.5°C days at a humidity of 70–80 per cent), $T_{min,bd}$ is the minimum temperature required for the digestion of the blood meal (9.9 °C) and T is the actual average temperature (ranging between $T_{min,bd}$ and the maximum temperature of about 40°C). The above relationship, estimated for *A. maculipennis s.l.*, is used because it agrees fairly well with observations of other *Anopheles* (Boyd, 1949; Macdonald, 1957; Gillies & De Meillon, 1968; Garrett-Jones & Shidrawi, 1969; White, 1982).

The female mosquito has to live long enough for the parasite to complete its development if transmission is to occur. Longevity of the mosquito vector depends mainly on the vector species, humidity, the availability of hosts, and temperature. A number of survival probabilities for various *Anopheles* species are given in

Table 3.3: Values of Daily Survival Probabilities of Several Anophelines

Anopheles species	*p*	Reference
A. albimanus	0.65–0.91[a]	Weidhaas et al. (1974)
A. antroparvus	0.85	Horsfall (1955)
A. coustani	0.89	Garrett-Jones & Grab (1964)
A. culcifaries	0.94	de Zoysa et al. (1988)
A. freeborni	0.97	Horsfall (1955)
A. freeborni	0.72–0.75	McHugh (1989)
A. funestus	0.91	Garrett-Jones & Grab (1964)
A. funestus	0.89	Service (1965)
A. funestus	0.89	Garrett-Jones & Shidrawi (1969)
A. funestus	0.82–0.87	Molineaux & Gramiccia (1980)
A. gambiae	0.91–0.93	Davidson (1954)
A. gambiae	0.88	Garrett-Jones & Grab (1964)
A. gambiae	0.90	Service (1965)
A. gambiae	0.83–0.88	Garrett-Jones & Shidrawi (1969)
A. gambiae	0.76–0.94[b]	Zahar (1974)
A. gambiae	0.82–0.87	Molineaux & Gramiccia (1980)
A. jeyporiensis	0.81	Khan & Talibi (1972)
A. messae	0.81–0.85	Horsfall (1955)
A. minimus	0.90	Khan & Talibi (1972)
A. nili	0.84	Garrett-Jones & Grab (1964)
A. pharoensis	0.62[b]	Zahar (1974)
A. philippinensis	0.72	Khan & Talibi (1972)
A. quadrimaculatus	0.97[d]	Horsfall (1955)
A. sacharovi	0.64–0.85[c]	Horsfall (1955)
A. sinensis	0.96[d]	Horsfall (1955)
A. stephensi	0.83–0.87	Zahar (1974)
A. vagus	0.96[d]	Horsfall (1955)
A. vagus	0.80	Khan & Talibi (1972)

[a] Wet season *p* between 0.73–0.91;
   dry season *p* between 0.65–0.70.
[b] Under insecticide spraying.
[c] Low values according to low relative humidity.
[d] Bred in cages.

Table 3.3. Between certain temperature thresholds, the longevity of a mosquito decreases with rising temperature (Molineaux, 1988). The optimum temperature for mosquito survival lies in the 20°C–25°C range. Temperatures in excess of these will increase mortality and there is a threshold temperature above which rapid death is inevitable. By the same token, there is a minimum temperature below which the mosquito cannot become active. Relying on data reported by Boyd (1949), Horsfall (1955), and Clements & Paterson (1981), we assume a daily survival probability of 0.82, 0.90 and 0.04 at temperatures of 9°C, 20°C and 40°C, respectively (see Figure 3.3C), expressed as:

$$p = e^{-1/(-4.4+1.31T-0.03T^2)}$$ (3.6)

The relative humidity is assumed to remain at a favourable level for mosquito development and not to change with changing precipitation.

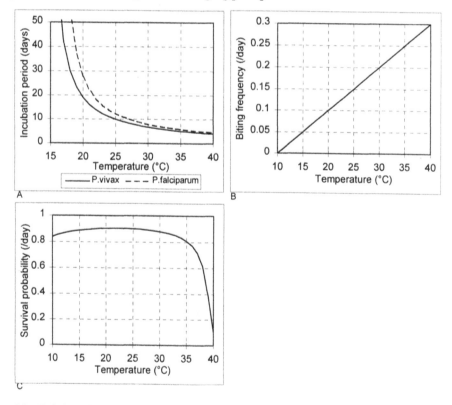

Figure 3.3: Relationship between Temperature and Factors of Malaria Transmission. (A): Incubation Period of Malaria Parasites in the Mosquito; (B): Mosquito Biting Frequency; (C): Survival Probability of Adult Mosquito

Dengue

In the case of dengue the relationship between temperature and model parameters is based on the well validated mosquito simulation model (CIMSiM) and the accompanying dengue transmission model (DENSiM) (Focks *et al.*, 1993a, b, 1995). These models have combined the influence of temperature on adult mosquito survival, the length of the gonotrophic cycle, the EIP and the number of replete feeds required for a gonotrophic cycle, and have accurately predicted dengue transmission, validated against historical data (Focks *et al.*, 1995).

A detailed enzyme kinetic model describes the relation between temperature and the incubation period of dengue *Flavivirus*, and that between temperature and blood digestion. This model assumes that, if other factors are not limiting, the rate of development is determined by a single rate-controlling enzyme, which is denatured reversibly at high and low temperatures:

$$
r = \frac{P_{(25°C)} \dfrac{T}{298} e^{\frac{\Delta H_A^\neq}{R}\left(\frac{1}{298}-\frac{1}{T}\right)}}{1+e^{\frac{\Delta H_H^\neq}{R}\left(\frac{1}{T_{1/2H}}-\frac{1}{T}\right)}}
\tag{3.7}
$$

where r represents the development rate (in hours) at temperature T (in K), $P_{(25°C)}$ is the development rate at 25°C, assuming no inactivation of the critical enzyme, $\Delta H_A^\neq$ is the enthalpy of activation of the reaction that is catalysed by the enzyme, $\Delta H_H^\neq$ is the enthalpy change associated with high temperature inactivation of the enzyme, $T_{1/2H}$ is the temperature (K) at which 50 per cent of the enzyme is inactivated at high-temperature, and R is the universal gas constant (= 1.987 cal/mol/deg). The statistical methods used to estimate parameter values can be found in Focks *et al.* (1993a, b). Parameter values for the enzyme kinetics model for the calculation of the development rate of the dengue virus and the rate of digestion of a blood meal are: $P_{(25°C)}$=0.003359 and 0.00898 per hour; $\Delta H_A^\neq =$ 15,000 and 15,725.23 cal/mol; $\Delta H_H^\neq = 6.203E30$ and 1,756,481.07 cal/mol; $T_{1/2H}$ = -2.176E30 and 447.17 K, respectively.

The incubation period (n) is completed as soon as the cumulative development exceeds 1. In general, dengue viruses do not survive below about 12°C–13°C, and a minimum temperature for virus development of 11.9°C is used. Watts *et al.* (1987) reported a relationship between the virus titer in the infecting blood meal and the subsequent duration of the EIP. The scaling factor (as a function of the virus titer) for duration of the EIP is set to 1, corresponding to a virus titer of approximately 5.4 MID$_{50}$ (Focks *et al.*, 1995). Figure 3.4A depicts the relationship between temperature and incubation period, resulting from the above description.

The first gonotrophic cycle is completed when the cumulative development rate exceeds 1. Subsequent cycles are completed when the cumulative rate has increased by an additional 0.58. In contradistinction to anophelines, multiple feeding is very common in *Aedes* species (McClelland & Conway, 1971; Scott *et al.*, 1993a). The number of blood meals taken during a gonotrophic cycle depends on the female's weight, which in turn is temperature-dependent. At warmer temperatures smaller adults hatch that often require multiple feeds to develop their eggs. The relationship between female weight and temperature is based on Rueda *et al.* (1990); the number of complete, replete feeds decreases linearly from 2 at a female weight less than 0.5 mg to 1.1 at female weights above 3.5 mg (Focks *et al.*, 1993a).

Another factor contributing to the feeding frequency is the habit of *Aedes* mosquitoes of feeding or probing on a number of people within the same room. Assuming four feeding attempts per meal and a probability of feeding on different human hosts of 0.6, the average number of different people probed or fed upon is set to 2.8 per replete feed (Focks *et al.*, 1995).

Aedes aegypti is usually considered a rather selective human feeder although it is known to feed on other hosts, too (Christophers, 1960). The value of 90 per cent human blood meals (HBI) used in the model is probably reasonable for *A. aegypti* in many tropical regions (Scott *et al.*, 1993b). Figure 3.4B shows the relationship between temperature and man-biting habit (*a*), taking into account the factors described above.

Aedes mosquitoes are less responsive to ambient temperatures than anophelines since they live mainly indoors. The relationship between the survival probability (*p*) and temperature used is based on numerous sources and is described in Focks *et al.* (1993a): the survival is set at 0.89 per day in temperature ranges between 6°C and 40°C. Below or above these temperatures, survival declines progressively (see Figure 3.4C) .

As is the case with malaria transmission, there is no evidence of climate affecting the parameters *b* and *c* in the EP equation. There is evidence that the titer of virus in the blood meal could influence transfer probabilities (e.g. Gubler, 1976, 1987). A reasonable generic value of approximately 5.4 MID_{50} is used (Focks *et al.*, 1995), corresponding to an infection probability after completion of the EIP of approximately 0.45 (Watts *et al.*, 1987; Newton & Reiter, 1992). Human susceptibility is set to 0.9.

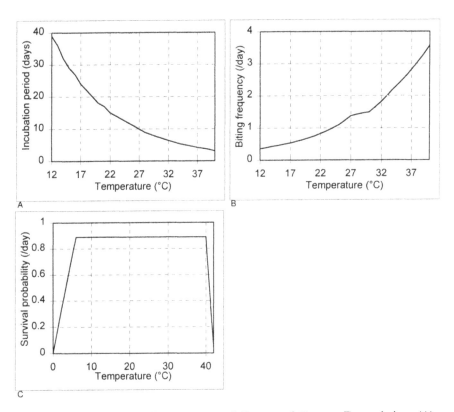

Figure 3.4: Relationship between Temperature and Factors of Dengue Transmission. (A): Incubation Period of Dengue Virus in the Mosquito; (B): Mosquito Biting Frequency; (C): Survival Probability of Adult Mosquito

Schistosomiasis

Miracidia and cercariae are short-lived, non-feeding organisms which have relatively large glycogen reserves. The duration of the viability of these larval stages depends on the quantity of these storehouses of energy. Extrinsic factors, especially temperature, which stimulate the use of the glycogen curtail larval viability (Jordan & Webbe, 1982). *S. mansoni, S. japonicum* and *S. haematobium* miracidia respond to different temperatures in a very similar manner, surviving longer at moderate temperatures than at high or low ones. Optimum temperatures for survival fluctuate around 15°C (Prah & James, 1977). It is assumed that the same relationship between temperature and life expectancy applies in the case of cercariae. As they generally live longer, their life expectancy is 1.5 times the life expectancy of the miracidia. The death rate in the larval stages is 1 divided by the life expectancy (Anderson *et al.*, 1982) (see Figure 3.5A).

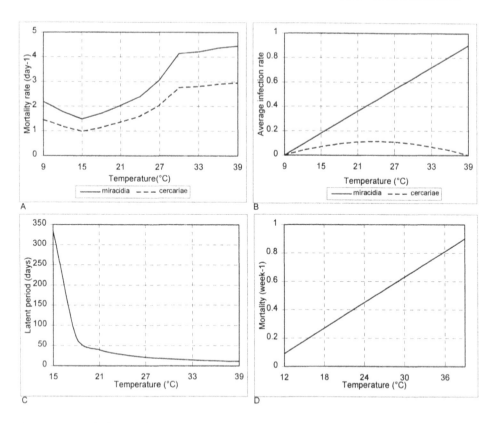

Figure 3.5: Relationship between Temperature and Factors of Schistosomiasis Transmission. (A): Life Expectancy of the Miracidia and Cercariae Life Expectancy; (B): Average Infection Rate of the Miracidia and Cercariae; (C): Incubation Period of the Schistosomes in the Snail; (D): Per Capita Adult Snail Mortality Rate

The effect of temperature on the ability of cercariae to infect humans is not identical to its effect on miracidia in infecting snails. At low temperatures both miracidia and cercariae move slowly, so the chance of infection during the larval stages is low. As temperature increases, both snail and miracidia become more active, leading to more favourable conditions for contact between the miracidia and the snail. The infection rates for all three species of miracidia increase as temperature rises, while infection does not occur at all at temperatures below 9°C. The cercariae achieve a maximum infection rate at 24°C–27°C. A linear function describes the relation between the infection rates achieved by miracidia and temperatures up to the thermal death point of the snails (39°C). In the case of cercariae, the infection rate increases curvilinearly to a peak of 24°C and then decreases symmetrically (Purnell, 1966) (Figure 3.5B).

The incubation period of the schistosomes in the snail depends critically on temperature. When temperatures are rising, larval development rate increases for all three parasite species, independent of the species of snail in which larval development occurs (Prah & James, 1977). Below approximately 15°C larval development is in general completely inhibited, while above 39°C snail and parasite death begin to occur rapidly. As is the case for malaria, the relation between temperature and the latent period of the parasite inside the snail can be expressed as a temperature sum (Plüger, 1980):

$$n = \frac{D_{st}}{T - T_{min,st}} \qquad (3.8)$$

where n is the latent period of the schistosomes inside the snail (days), D_{st} the number of degree days needed for schistosome development inside the snail (= 268 °C days), T the water temperature, and $T_{min,st}$ the minimum temperature needed for parasite development (= 14.2°C) (Figure 3.5C).

Temperature affects the complex relationship between snail mortality and infection rate. Infection tends to reduce the life expectancy of the snail by a factor of 3, so the change in infection rate as temperature fluctuates indirectly determines snail mortality. Temperature also directly influences snail mortality, which increases as temperature rises. A linear relationship is used between the per capita adult snail mortality rate and mean water temperature (Anderson & May, 1979; Woolhouse & Chandiwana, 1990):

$$\mu_{ss} = b + dT \qquad (3.9)$$

where μ_{SS} is the death rate of the shedding snails, b (= -0.374) and d (=0.0329) are constants, and T is the mean water temperature (Figure 3.5D).

Rainfall and Vegetation Patterns

Rainfall, or the lack of it, plays a crucial role in malaria and schistosomiasis epidemiology. Rainfall not only provides the medium for the aquatic stages of the malarial mosquito's life cycle, but also may increase the relative humidity and hence increase the longevity of the adult mosquito. In schistosomiasis epidemiology, rainfall determines the duration of desiccation, and the snails are affected by fluctuations in the water level as well as by the rapid increase in the velocity of flows of water after heavy rains (Sturrock, 1973). The relationships between changing temperatures, precipitation and relative humidity, however, are complicated, and the processes affecting atmospheric humidity suggest only a small change in relative humidity as the atmosphere gets warmer due to the greenhouse effect (Mitchell & Ingram, 1992). The introduction of large-scale irrigation schemes has also reduced the significance of local rainfall in vector-

---

Box 3.2:
Potential Evotranspiration

The combined evaporation of the soil surface and transpiration from plants, called 'evotranspiration', represents transport of water from the Earth back to the atmosphere. Potential evotranspiration (pet) is calculated as a function of monthly temperature, T (in °C) (Thornthwaite, 1948; Mather, 1978):

$$\text{pet} = 16c(10T / I)^d \tag{3.10}$$

where c is an empirically derived adjustment for the length of the day and latitude of the region; d is a non-linear function of the annual heat index (I), which is the sum of 12 monthly heat indices i:

$$d = 0.000000675\, I^3 - 0.00007711\, I^2 + 0.01792 I + 0.49239 \tag{3.11}$$

$$i = (T / 5)^{1.514} \tag{3.12}$$

The mathematical derivation of these formulas is empirical and the formulas – although lacking mathematical elegance – allow calculation of the potential evotranspiration with only a minimum amount of input data. However, over tall vegetation, potential evotranspiration may be larger than from open water. In this case, potential evotranspiration is overestimated by the Thornthwaite–Mather equations as described above. Furthermore, evotranspiration depends also on water availability at the ground surface and the moisture deficit in the air above the ground. Using climate change scenarios in combination with the 'temperature driven' Thornthwaite–Mather equations, may again overestimate the potential evotranspiration (Martin & Lefebvre, 1995).

---

borne disease epidemiology to some extent (Muir, 1988).

The ability of the malaria and schistosomiasis vectors to reproduce depends more or less on whether they encounter motionless or rapidly moving water and on the type of vegetation. Following Martin & Lefebvre (1995), the ratio of precipitation to potential evotranspiration (the moisture index) is used to describe savannahs, which are among the driest habitats (see Box 3.2). A value of 0.6 for the moisture index is used to describe savannahs (Box, 1981); below this moisture index transmission is assumed not to occur (Kloos & Thompson, 1979; Martin & Lefebvre, 1995). These areas are therefore excluded from the model simulations of potential transmission areas, although incidental transmission may occur there (e.g. in the neighbourhood of an oasis).

As *A. aegypti* is mainly found indoors or in domestic habitats, rainfall plays less of a role in dengue epidemiology than it does in malaria and schistosomiasis epidemiology. Although rainfall does influence the availability of artificial and natural breeding containers in some locations (for example in Puerto Rico (Moore *et al.*, 1978)), in many places the number of breeding sites does not vary with rainfall (e.g. in Bangkok (Southwood *et al.*, 1972)). Furthermore, because the mosquitoes live indoors, adult survival does not correlate seasonally with humidity. The moisture index described above is therefore not used in the assessment of the EP of dengue.

MALARIA PREVALENCE

The history of a mathematical approach to malaria is nearly as old as the discovery of its transmission dynamics. The earliest attempt to arrive at a quantitative understanding of the dynamics of malaria transmission was made by Ross (1911), whose models consist of a handful of differential equations that describe changes in the densities of susceptible and infected people and mosquitoes. In the 1950s Macdonald (1957) added a dimension of biological realism to these early models by his careful attention to the interpretation and estimation of parameters. Although these basic models provide a useful overview of the dynamics of malarial infection, many of their predictions deviate strikingly from reality.

An obvious modification to the basic model would be the incorporation of latent periods during which hosts are infected but not yet capable of transmitting the disease. Furthermore, models addressing the transmission dynamics of malaria have begun to take account of the phenomenon of acquired immunity. The reason for this belated attention to immunity development is in part a consequence of the early focus in malaria models on the vector component in transmission, which is in turn explained by the initial aim of global eradication of malaria by means of the application of dichloro-diphenyl-trichloro-ethane (DDT). Aron & May (1982) have since provided a simple way to incorporate the observed mechanism of the maintenance of immunity with continuous exposure. Although their model represents an advance over the earlier simple models, it nevertheless remains a very crude approximation of the true complexities of immunity to malarial infection.

The model used to describe the transition between the reservoirs of the human population at risk is based on a microparasite–epidemiological model, described in Aron & May (1982), Bailey (1982), Levin *et al.* (1989) and Anderson & May (1991) (see Box 3.3). However, two important aspects are ignored in this framework (Anderson & May, 1991). First, in this model no distinction is drawn between infection and infectiousness. An infected person may harbour the parasite in certain stages of its development (in liver and blood) but not harbour parasites at a stage which is infective to the vector. Neither is the intensity of the infection described in the model (since only the presence or absence of infection is modelled). The second shortcoming concerns the nature of acquired immunity to

Box 3.3:
An Epidemiological Malaria Model

The human population subject to a risk of malaria, for a given age a, is divided into three categories: susceptible persons (X), infected persons (Y), and immune persons (Z) (see Figure 4.2). The latent reservoir is omitted, because the duration of a stay in this reservoir is usually very short in comparison with the residence time in the other reservoirs. $\lambda(t)$, $\gamma(t)$ and $v(t)$ are defined below.

$$\frac{\delta X}{\delta t} + \frac{\delta X}{\delta a} = \gamma(t)Z(a,t) - (\lambda(t) + \mu)X(a,t) \tag{3.13}$$

$$\frac{\delta Y}{\delta t} + \frac{\delta Y}{\delta a} = \lambda(t)X(a,t) - (v(t) + \mu)Y(a,t) \tag{3.14}$$

$$\frac{\delta Z}{\delta t} + \frac{\delta Z}{\delta a} = v(t)Y(a,t) - (\gamma(t) + \mu)Z(a,t) \tag{3.15}$$

$$\lambda(t) = VC \times Y'(t) \tag{3.16}$$

$$\gamma(t) = \frac{\lambda(t)}{e^{\lambda(t)\tau} - 1} \tag{3.17}$$

$$v(t) = \frac{\lambda(t)}{e^{\lambda(t)v} - 1} \tag{3.18}$$

The number of susceptible persons may change over time, as they become members of the infected class at a rate λ. Infected individuals may recover to join the immune class (at a rate v). Immune people lose their immunity at a rate γ, and those who have lost their immunity return to the reservoir of susceptibles. All new-born babies are assumed to be members of the class of susceptibles, and as they grow older they graduate from one age class to the next (age classes of 1 year). The natural death rate is denoted by μ, and, for simplicity's sake, equals the birth rate (0.02/year). The rate at which individuals become infected (λ) depends on the vectorial capacity (VC), which represents the transmission potential of the mosquito population, and on proportion of infected people in the human population (Y). VC is expressed as:

$$VC = k_1 \frac{a^2 p^n}{-\ln(p)} \tag{3.19}$$

where k_1 is a calibration coefficient, comparable to the term c_1 in equation 3.1, except that in k_1 the mosquito density is incorporated (which is assumed to be constant, see the section on the EP). The other variables are as in equation 3.1. The average infectious and immune periods are $1/v$ and $1/\gamma$, respectively. Assuming that re-exposure does not occur, infection and immunity endure for a fixed period of time, denoted by v and τ (average 3 years infected with P. vivax and 1 year with P. falciparum when left untreated (Molineaux, 1988); the basic loss rate of immunity is 0.67/year, corresponding with a mean duration of immunity of 1.5 years (Aron & May, 1982)). However, if a person is further exposed before this period has elapsed, infection and immunity are sustained and the average rate of loss of infection or immunity can be described by the formulas given above. Thus, given a certain pattern of VC, we can use the equations set out above to simulate the transmission dynamics in a given population.

malarial infection. Shaking off the infection does not imply full protective immunity against re-infection. To speak of a single reservoir of immune persons is an oversimplification of the true complexities of acquired immunity to malaria.

CLIMATE SCENARIOS

The climate changes referred to in the following sections were constructed by the IPCC Working Group II: Impacts Assessment (Viner, 1994). Documented current climate conditions were modelled according to the results of three transient GCMs: the GCM developed by the Max Planck Institute in Germany (ECHAM1-A) (Cubasch *et al.*, 1992), the GCM developed by the UK Meteorological Office (UKTR) (Murphy, 1995; Murphy & Mitchell, 1995) and the GCM developed at the Geophysical Fluid Dynamics Laboratory in the USA (GFDL89) (Manabe *et al.*, 1991, 1992). The baseline climatology relies upon temperature data for the period 1951–1980 (Legates & Willmott, 1990a, b). The value for the global mean temperature change of 1.16°C, which lies close to the IPCC 'best estimate' around the year 2050, is used to identify the decades in the three transient GCM experiments where the global mean temperature is equivalent to this value of 1.16°C (Viner, 1994). The climate scenarios derived from these GCM experiments must be regarded not as future climate predictions, but merely as one possible interpretation of how global climate may evolve in the future. Therefore, although the average global temperature increase is the same in the three experiments, the patterns of climate change are different.

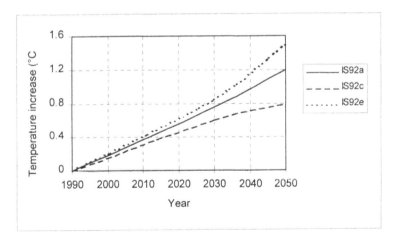

Figure 3.6: Global Temperature Changes According to the IS92a, IS92c and IS92e Scenarios

Using the 'simple linked method', described in Viner (1994), incremental global mean temperature changes are projected on the baseline climate by standardising the output of the three GCMs. Global mean temperature changes and population growth of three IPCC scenarios are used: the IS92a, IS92c, and IS92e scenarios (Leggett *et al.*, 1992). The IS92a and IS92e scenarios adopt the World Bank (1991) population estimates; the population estimate in the IS92c scenario is based on the United Nations (UN) medium low case scenario (UN, 1992). Figure 3.6 shows the global average temperature increase associated with these three scenarios.

CHANGES IN POTENTIAL RISK AREAS

Different populations will respond differently to any specific changes in transmission rates caused by climatic changes, and prevalence will depend to a significant extent on other, more local, factors. Nevertheless, in this section an attempt is made to generate a number of plausible projections of changes in the numbers of people at risk and infected as climate changes.

The EP as defined in equations 3.1 and 3.2 is used as a comparative index to estimate the effect on malaria, dengue, and schistosomiasis transmission potential of a change in average monthly temperature and precipitation patterns, estimated using the three GCMs. Given the difference in minimum temperature for parasite development, a distinction is made between the *P. vivax* and *P. falciparum* parasites; the various species of dengue viruses and schistosomes were considered as a group. Since no reliable field estimates of c_1 and c_2 are available, the *absolute* values of the EP should be treated with caution; more important are the *relative* differences (the trends) in EP over time.

Combining the temperature effects on the transmission cycles of malaria, dengue and schistosomiasis, described before, Figure 3.7 shows that at high temperatures, in the case of malaria and dengue, the accelerated development of the parasite and the increased biting rate can no longer compensate for the decreasing mean life expectancy among the mosquitoes. In the case of schistosome species the increase in EP associated with the increasing infective potency of the miracidia and the decrease in the latent period of the schistosome parasite at higher temperatures is cancelled by the increase in mortality rates among miracidia, cercariae and snails. Maximum values for EP, i.e. transmission potential given a certain vector density, are found in the ranges 29°C–33°C, 16°C–18°C and 40°C for malaria, schistosomiasis and dengue respectively. Varying the most sensitive parameters in the EP equation, namely host mortality rates and development time of the parasite, does not strongly influence the pattern of change. Moreover, the actual transmission intensity also depends on vector abundance, so that within the optimum temperature range for vector breeding and reproduction (20°C–30°C) and

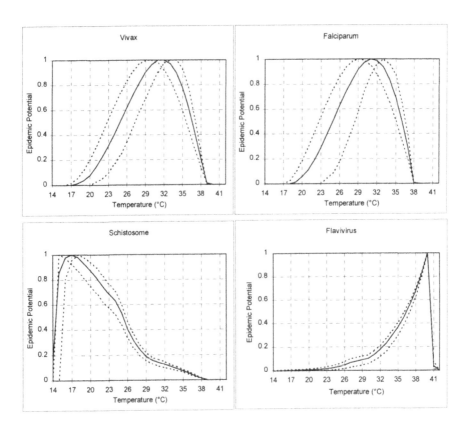

Figure 3.7: EP (Valued as 1 as a Maximum) for *P. vivax* (Left-Hand Curve $p(20°C) = 0.95$ and $T_{min,m} = 14.5°C$; Central Estimate $p(20°C) = 0.9$ and $T_{min,m} = 14.5°C$; Right-Hand Curve $p(20°C) = 0.8$ and $T_{min,m} = 15°C$), *P. falciparum* (Left-Hand Curve $p(20°C) = 0.95$ and $T_{min,m} = 16°C$; Central Estimate $p(20°C) = 0.9$ and $T_{min,m} = 16°C$; Right-Hand Curve $p(20°C) = 0.8$ and $T_{min,m} = 19°C$) and Schistosome Parasites (Left-Hand Curve $\mu_{ss}(25°C) = 0.1/week$ (Chu *et al.*, 1966) and $T_{min,st} = 14°C$; Central Estimate $\mu_{ss}(25°C) = 0.44/week$ (Sturrock & Webbe, 1971) and $T_{min,st} = 14.2°C$; Right-Hand Curve $\mu_{ss}(25°C) = 0.6/week$ and $T_{min,st} = 15°C$) and *Flavivirus* (Left-Hand Curve $p(6–40°C) = 0.96$; Central Estimate $p(6–40°C) = 0.89$; Right-Hand Curve $p(6–40°C) = 0.76$ (Sheppard *et al.*, 1969))

where, in addition, rainfall and humidity are optimum, our results probably underestimate the actual change in transmission potential of malaria, dengue and schistosomiasis vector populations. For example, the optimum temperature range for schistosomiasis transmission in parts of Africa takes place with water temperatures of about 22°C–27°C (Shiff, 1964; Shiff *et al.*, 1979).

Figure 3.8A–D depicts the limits of the current potential geographical extent of malaria, dengue, and schistosomiasis transmission, and the global distribution of the potential risk areas arrived at using the average temperature and precipitation

changes of the GFDL, ECHAM, and UKTR GCMs (to average the results of the
GCMs may be disputed; therefore, the figures are only meant to illustrate the
change in distribution limits). In the case of *P. vivax*, this includes large areas in
the USA up to the Canadian border, southern and central Europe, Turkey, southern
Russia, China and Japan. *P. falciparum* malaria is restricted to more tropical areas
since parasite development needs a minimum temperature of approximately 16°C.
As dengue viruses generally do not survive below mean temperatures of about
12°C–13°C, and no moisture limitation is used for its transmission, this vector-
borne disease has a more extended potential distribution than malaria and
schistosomiasis. Besides the current endemic schistosomiasis areas in Africa,
South-East Asia and South America, simulated current potential schistosomiasis
areas likewise involve Europe, North America and large regions of Asia.

A comparison of the potential geographical extent of the three diseases with the
actual distribution indicates that the simulation of future risk areas must be
interpreted to take account of local conditions and developments. In tropical and
subtropical regions, climatic conditions are already favourable for vector breeding
and reproduction, resulting in vector densities which exceed the critical value
during large parts of the year. However, in other regions disease transmission is
absent, although climatic factors, particularly temperature, are apparently
conducive to transmission. Therefore, discrepancies between observed and
expected values of EP may be due to either effective vector control measures, the
treatment of infected individuals, or specific characteristics of the human and/or
vector population.

Figure 3.8 shows that an expansion of the geographical areas susceptible to
malaria and schistosomiasis transmission is to be expected as the climate changes.
The main changes, relative to the baseline climate, would occur in the areas with
temperate climates in which vectors capable of transmission already occur, but
development of the parasite has hitherto been inhibited by temperature.
Simulations run on the three GCMs (Figure 3.9A–D) show that large parts of
North America, Europe, Australia and Asia would be at risk of malaria, dengue
and schistosomiasis transmission even if a vector density 2 or more times lower
than the current level were to obtain.

Table 3.4: Estimated Relative Change (Per Cent) of the EP (ΔEP) in the Baseline Potential Risk
Areas in Africa, South-East Asia and South and Central America; the Climate Change Scenarios
Used Are the GFDL89, UKTR and ECHAM1-A Scenarios

GCM	*P. vivax*	*P. falciparum*	Schistosome	*Flavivirus*
GFDL89	23 (15–39)	27 (16–74)	-12 (-13—12)	45 (35–69)
UKTR	12 (7–23)	15 (8–45)	-11 (-11—10)	31 (24–47)
ECHAM1-A	17 (10–31)	20 (11–60)	-17 (-18—17)	47 (37–74)

The sensitivity studies underpinning Figures 3.8 and 3.9 also indicate that an increase in the EP of the malarial and dengue parasite in the presently vulnerable regions in the (sub)tropics is to be anticipated (Table 3.4) for all three climate change scenarios. On aggregate, this varies between ~10 and 75 per cent for malaria and between ~25 and 75 per cent for dengue. Thus, in areas in which malaria and dengue were previously only transmitted during certain months of the year, they may become a year long threat. As a consequence of the relatively low optimum transmission temperature given a certain snail density, the transmission potential of schistosomiasis may decrease between ~10 and 20 per cent in the present potential areas. However, in the current highly endemic areas, the prevalence of infection is persistently high, and will probably only be marginally affected by these climate-induced changes.

It is hardly possible to foresee whether the *potential* risk areas, as simulated in Figure 3.8, are likely to experience epidemics or become endemic in the future. Therefore it is assumed that the present endemic areas will be supplemented by areas in which the simulated annual average EP doubles, or increases by a factor greater than 5 (a doubling of the EP or an increase of the EP by a factor 5 means that the vector density needed to establish an endemic situation decreases by a factor 2 or by a factor 5, respectively). Although the transmission potential of malaria, dengue and schistosomiasis in some of the present endemic areas may decrease, transmission rates will probably remain high enough for endemicity to remain established. Only the changes in EP in less economically-developed areas are considered, as most developed countries will be in a position to take mitigating measures as transmission potential increases.

Numbers of people in the developing world at risk of contracting malaria, schistosomiasis, and dengue may change between ~5 and ~15 per cent, ~0 and ~5 per cent, and ~0 and ~5 per cent, respectively, depending on the GCM climate scenario. Assuming that the population in the developing world increases up to a total of ~8.6 billion by the year 2050 according to the IS92a scenario, the year corresponding to a global average temperature increase of 1.16°C (Viner, 1994), and that the population is evenly distributed in the areas under consideration, the *additional* number of people at risk due to anthropogenic climate change can be estimated. Around 2050 this may increase up to ~720 million, ~40 million, and ~195 million people with malaria, schistosomiasis, and dengue respectively. Table 3.5 summarises the change in the population at risk of contracting malaria, dengue and schistosomiasis as simulated with the model.

P Vivax

(a) Baseline climate

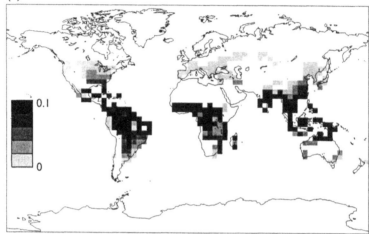

(b) Climate change scenario

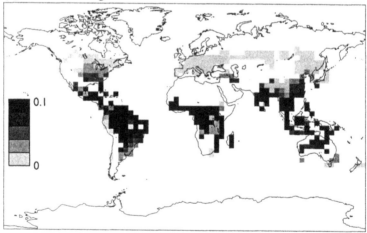

Figure 3.8A: Potential Risk Areas for Baseline Climate (1931–1980) and Climate Change Scenario; Based on the Average Climate Change Patterns Generated by the GFDL89, UKTR and ECHAM1-A GCMs for *P. vivax*, Calculated from Monthly Temperature and Precipitation. Global Mean Temperature Increase According to the Three Scenarios is 1.16°C

P Falciparum

(a) Baseline climate

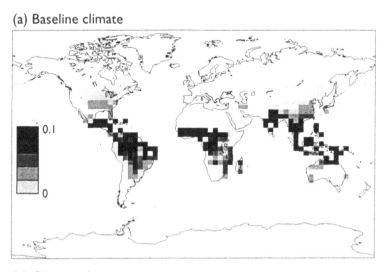

(b) Climate change scenario

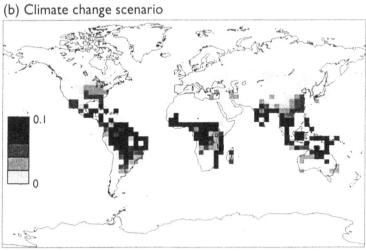

Figure 3.8B: Potential Risk Areas for Baseline Climate and Climate Change Scenario for *P. falciparum* (see Caption Figure 3.8A)

Schistosomiasis

(a) Baseline climate

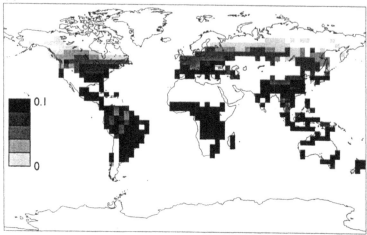

(b) Climate change scenario

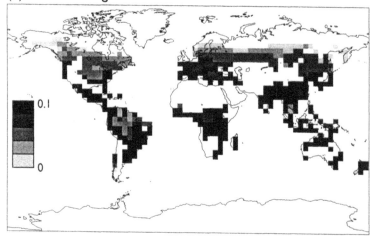

Figure 3.8C: Potential Risk Areas for Baseline Climate and Climate Change Scenario for *Schistosomes* (See Caption Figure 3.8A)

Dengue

(a) Baseline climate

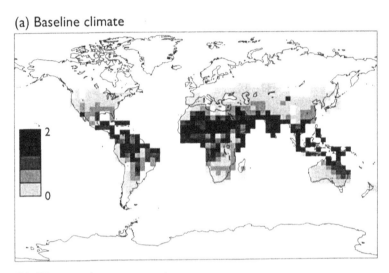

(b) Climate change scenario

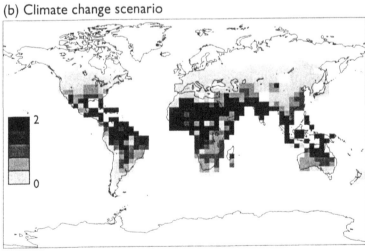

Figure 3.8D: Potential Risk Areas for Baseline Climate and Climate Change Scenario for Dengue Viruses (See Caption Figure 3.8A)

P Vivax

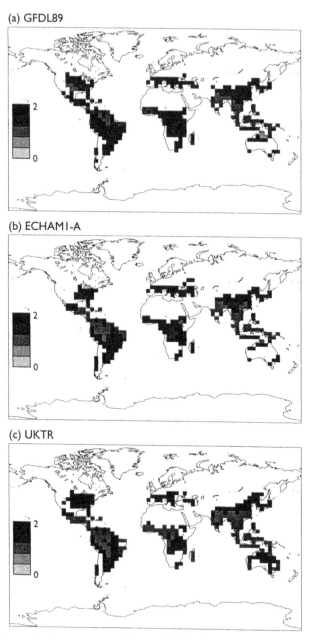

(a) GFDL89

(b) ECHAM1-A

(c) UKTR

Figure 3.9A: Changes in Average Annual Potential Risk Relative to Baseline Climate Based on the Climate Patterns Generated by the GFDL89, ECHAM1-A and UKTR GCMs for *P. vivax*, Calculated from Monthly Temperature and Precipitation; Global Mean Temperature Increase According to the Three Scenarios is 1.16°C; in the Representation on the Maps, an Arbitrary Cut-Off Value for Current EP of 0.01 for Malaria is Used; This is to Reduce Bias towards Large Changes that Would Result from Using Infinitesimal EP Values as Denominators in Regions Currently at Near-Zero Risk of the Disease

P Falciparum

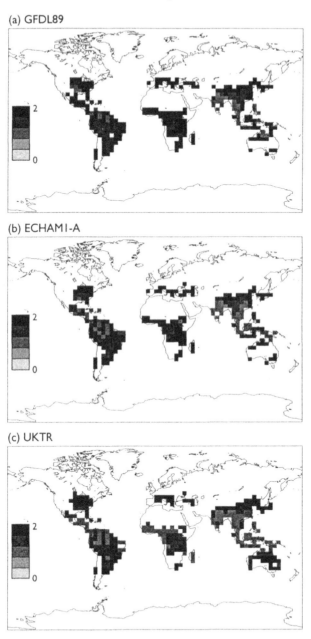

Figure 3.9B: Changes in Average Annual Potential Risk Relative to Baseline Climate Based on the Climate Patterns Generated by the GFDL89, ECHAM1-A and UKTR GCMs for *P. falciparum* (See Caption Figure 3.9A)

Schistosomiasis

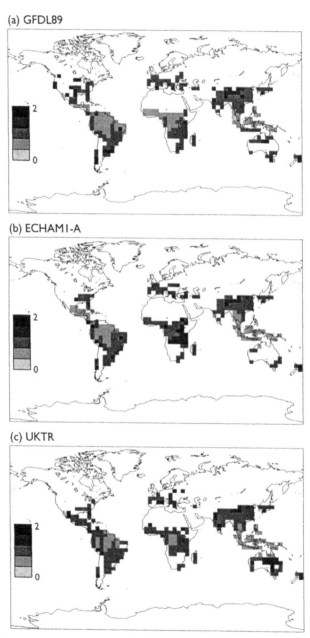

(a) GFDL89

(b) ECHAM1-A

(c) UKTR

Figure 3.9C: Changes in Average Annual Potential Risk Relative to Baseline Climate Cased on the Climate Patterns Generated by the GFDL89, ECHAM1-A and UKTR GCMs for *Schistosomes* (see Caption Figure 3.9A); in the Representation on the Maps, an Arbitrary Cut-Off Value for Current EP of 0.08 is Used; This is to Reduce Bias Towards Large Changes that Would Result from Using Infinitesimal EP Values as Denominators in Regions Currently at Near-Zero Risk of These Diseases

Dengue

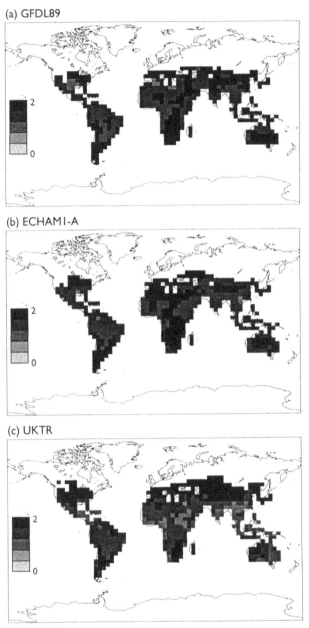

Figure 3.9 D: Changes in Average Annual Potential Risk Relative to Baseline Climate Based on the Climate Patterns Generated by the GFDL89, ECHAM1-A and UKTR GCMs for DEN Viruses (See Caption Figure 3.9A); in the Representation on the Maps, an Arbitrary Cut-Off Value for Current EP of 0.1 is Used; This is to Reduce Bias towards Large Changes that Would Result from Using Infinitesimal EP Values as Denominators in Regions Currently at Near-Zero Risk of These Diseases

Table 3.5: Change in Population at Risk of Contracting Malaria, Schistosomiasis and Dengue (Per Cent) (the Climate Change Scenarios Used Are the GFDL89, UKTR and ECHAM1-A Scenarios); a Change of, for Example, 10 Per Cent Means that the Proportion of the Population in the Developing World at Risk (which is Assumed to be ~60 Per Cent, ~15 Per Cent, and ~45 Per Cent in 1990 for Malaria, Schistosomiasis and Dengue, Respectively) Increases 1.1 Times

GCM		*P. vivax*	*P. falciparum*	Schistosomiasis	Dengue
GFDL89					
	>2 x EP	10 (8–23)	14 (9–57)	3 (3–4)	5 (1–35)
	>5 x EP	5 (4–7)	6 (5–23)	3 (3–3)	1 (0–1)
UKTR					
	> 2 x EP	8 (7–16)	11 (7–38)	3 (3–4)	5 (1–19)
	> 5 x EP	5 (4–7)	6 (5–19)	3 (3–3)	0 (0–1)
ECHAM1-A					
	>2 x EP	8 (5–19)	13 (7–46)	2 (4–2)	4 (0–37)
	>5 x EP	3 (2–5)	5 (5–21)	1 (1–2)	0 (0–0)

Tables 3.4 and 3.5 show that the change in vector-borne disease risk depends on the climate scenario used. However, as shown in Figure 3.7, the EP is sensitive to changes in vector survival rates as well as to parasite incubation times. Figure 3.10 shows the range in the percentage change in EP, as well as the range surrounding the change in population at risk estimates, using the same values for mosquito and snail survival and minimum parasite temperature as in Figure 3.7. Maximum *changes* in EP are simulated for the lowest survival probabilities for *Anopheles* and *Aedes* mosquitoes, simply explained from the definition of the EP. The lower the minimum temperature for parasite development, the less the relative increase of the EP. The change in risk population of malaria (*P. falciparum*) is most sensitive to climatic changes, followed by dengue and schistosomiasis. An explanation for the difference in estimate ranges is the different assumptions (on vector survival and parasite development) made to estimate the low and high limits of changes.

CHANGES IN MALARIA PREVALENCE

Although actual malaria prevalence and incidence figures are not very reliable in most endemic regions, a good estimate of the infection rate can be obtained from the rate of increase of prevalence with age in young children. As the malaria situation varies in terms of resistance to change, a distinction is drawn between areas of high endemicity, mainly found in tropical Africa, and areas of lower endemicity found in other parts of Africa, South America, and South-East Asia. In tropical Africa, attention has been restricted to *P. falciparum*, the predominant

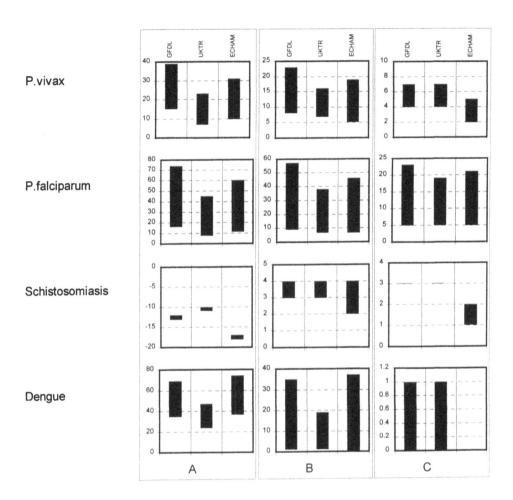

Figure 3.10: Range of the Estimated Relative Change (Per Cent) of the EP (ΔEP) in the Baseline Potential Risk Areas in Africa, South-East Asia and South and Central America (A), and the Range Surrounding the Estimates of Changes in Populations at Risk (B: EP increase >2x; C: EP increase >5x)

species responsible for most mortality. In regions of lower endemicity, changes in *P. vivax* and *P. falciparum* prevalence are simulated. The VCs (i.e. the averaged VC of the grids corresponding to the malarial regions) are chosen such that the force of infection is ~2.0 per year for the year 1990 in highly endemic regions and ~0.1 in areas of lower endemicity. Although these values are chosen rather arbitrarily, they lie within the range of the values found in several studies on the pristine force of infection in young children (Pull & Grab, 1974; Bekessy *et al.*, 1976). Choosing other initial values may alter the magnitude of change, but the direction of change remains the same. In the estimates of the excess prevalence in endemic areas, the malaria conditions are assumed to be in equilibrium in the year 1990. For the stable, highly endemic regions of tropical Africa, this assumption seems to be justified. However, for the unstable areas of lower endemicity this assumption will often be inappropriate.

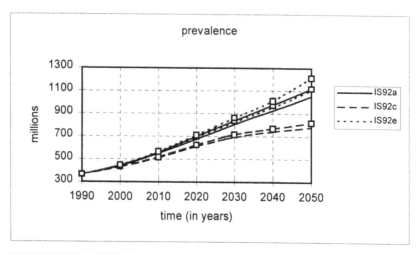

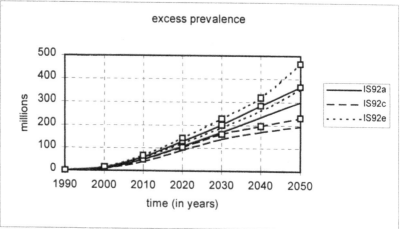

Figure 3.11: (Excess) Prevalence of Malaria Assuming that no Extension of the Current Regions Will Occur and that Present Endemic Areas Will be Supplemented by the Areas in which the EP Increases More than Twice (□)

Figure 3.11 shows the effect of a human-induced climate change on malaria prevalence. In the absence of climate change, an increase in prevalence is to be expected due to population growth. The simulations show an increase in malaria prevalence of ~320 million and ~170 million in 2050 due to population growth alone, for the IS92a, IS92e and IS92c scenario, respectively. If the wider areas become endemic, the effect of a human-induced climate change on the world prevalence of malaria will be more pronounced: excess prevalence due to climate change then varies between ~220 million and 480 million cases in the year 2050.

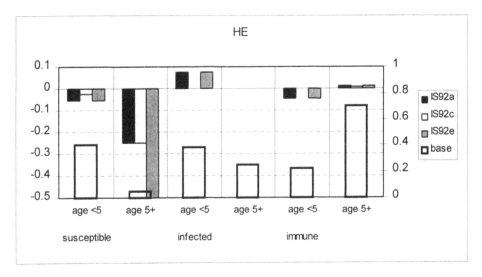

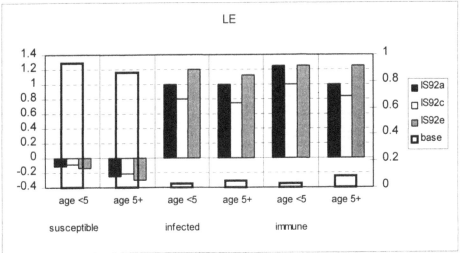

Figure 3.12: Changes in Susceptible, Infected and Immune People, for *P. falciparum* Malaria Regions of High Endemicity (HE) and Low Endemicity (LE); 'Base' Represents the Proportion of People Susceptible, Infected or Immune in the Initial Situation (Right *Y*-axis)

To illustrate the difference in malaria transmission dynamics, Figure 3.12 shows the change in people who are susceptible, infected, and immune, for *P. falciparum* for regions of high and low endemicity. In highly endemic areas, where the malaria situation is relatively stable, the result may be a decrease in numbers of people infected in the older age group as collective immunity increases. However, the increase in infected children under 5 is so pronounced that the prevalence per thousand population will increase. In areas of lower endemicity, a relatively small increase in malaria transmission potential may lead to a considerable increase in the prevalence of people suffering from malaria in both age groups, assuming that no special control measures are taken (e.g. increased application of antimalarial drugs). Although the increase in prevalence in the areas of lower endemicity is higher, the major part of the disease burden due to malaria remains in the highly endemic countries of tropical Africa. As *P. vivax* is considered not to be fatal, the disease burden due to infection with *P. vivax* is small compared to infection with *P. falciparum*.

LOCAL ESTIMATES: MODEL VALIDATION

It is difficult to validate the highly aggregated global model outcomes presented in the previous sections (see also Chapter 2). Although the underlying model relationships are conceptually valid, in that they reflect the prevailing theoretical insights into the part of reality that the model is supposed to represent, a lack of reliable historical global data sets makes pragmatic validation difficult. However, more confidence in the model outcomes can be obtained by validating the model on a local or regional scale, where data are at hand. An iterative cross-validation of large- and small-scale studies may be essential in the process of validating integrated assessment models (Root & Schneider, 1995). Below, model results are compared with local data regarding malaria and dengue transmission. No such attempt is yet made for schistosomiasis.

Malaria and Climate Change: A Case Study of Zimbabwe

In this section, we focus on the effect of climatic changes on malaria distribution in Zimbabwe. Zimbabwe, which lies at the southern fringes of malaria transmission in Africa (see Box 3.4) and has large regions with an altitude above 1500 m, is taken as an example. Zimbabwe is situated in south central Africa, is divided by a central watershed running from north-east to south-west, and lies in the southern limits of where malaria can occur in Africa. The country has a population of ~10 million people, of whom 45 per cent are below the age of 15 (World Bank, 1993). The vast majority of the population lives in the middle and high veld regions over 900 m altitude, and there are no defined migration patterns of direct relevance to

Box 3.4:
Highland Malaria in Africa
(Source: Lindsay & Martens (1998))

There is growing evidence that malaria is an increasing problem at the fringes of the current distribution limits. As appears from the previous sections, one of the causes of this rise in malaria may be a change in climatic conditions. In Africa the limits of transmission are often defined by altitude: higher altitudes may be too cold for transmission. Thus warming, due to short-term climate changes or global warming (or both), could push malaria transmission further up the slopes, resulting in epidemics in previously unexposed (and therefore immunologically naive) populations. Roughly, as altitude increase above 1500 m, malaria becomes less endemic (Nájera *et al.*, 1992). Figure 3.13 depicts regions in Africa above 1500 m.

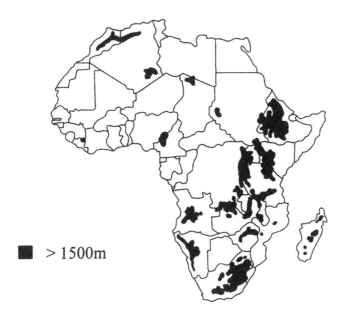

■ > 1500m

Figure 3.13: Highlands of Africa (Reproduced with the Kind Permission of S. Lindsay)

However, there are many other factors influencing the emergence of malaria, such as land use changes, population growth and movement, drug use and the development of resistance (see also Chapter 4), and the provision and utilisation of health services (Lindsay & Martens, 1997). For example, deforested areas may experience temperature increases of 3°C–4°C and the cleared land can often provide ideal breeding sites for some principal *Anopheles* vectors (Hamilton, 1989; Walsh *et al.*, 1993). In any case, most of these regions are particularly vulnerable due to a combination of the factors described above.

malaria (Taylor & Mutambu, 1986). Malaria is a major cause of morbidity and mortality in Zimbabwe, and is mainly attributable to infection with *P. falciparum*, with *A. arabiensis* as the major vector.

A recent study showed that temperature appears to have an effect on the severity of malaria (Freeman & Bradley, 1996). For the purposes of malaria epidemiology, Zimbabwe can be divided into seven clearly defined altitude zones (see Figure 3.14). The central watershed is defined by all land over 1200 m; decreasing in altitude to the south are the 900–1200 m south, 600–900 m south and <600 m south contours; to the north of the watershed, the same altitude zones may be distinguished: 900–1200 m north, 600–900 m north and <600 m north. In the areas below 600 m north, malaria is potentially perennial; in areas above 1200 m malaria normally does not occur; in other areas malaria is seasonal to epidemic in nature. The higher malaria transmission rates in the north and south below 600 m altitude result in a peak prevalence of *P. falciparum* in the age group of 5–9 years, not shown in the other malaria areas (Taylor & Mutambu, 1986).

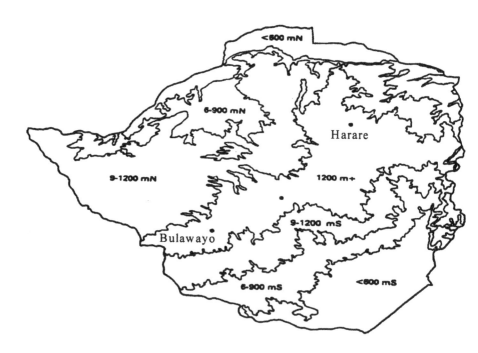

Figure 3.14: Altitudinal Classification of Zimbabwe (Reproduced with Permission from Taylor & Mutambu (1986))

The climate in Zimbabwe is hot and wet from January to April, cold and dry from May to August, and hot and dry from September to December. Malaria transmission occurs mainly during the rainy season, with most transmission occurring from February to May. During the winter months from June to August, temperatures are the coldest and transmission is extremely low or absent, whereas in the hot dry season, from August to October, much of the country is too dry due to the long dry season starting in May (Freeman, 1995). Transmission begins to increase with the onset of the rains in November (Taylor & Mutambu, 1986).

Figures 3.15 and 3.16 show the baseline EP as simulated with the EP model, for a baseline climate from 1931–1960 (Leemans & Cramer, 1991). For each altitude zone, grid cells corresponding to these zones are used to calculate the monthly EP. The pattern observed and described above is consistent with the model projections shown in Figure 3.15, taking into account a lag period of approximately 1 month between the peak in mosquito transmission potential and the peak of malaria occurrence. Figures 3.15 and 3.16 also clearly show the pattern of transmission intensity, which is highest at altitudes below 600 m at the north and south, and decreasing to the central higher altitudes (Taylor & Mutambu, 1986).

The baseline transmission potential is compared with a range of potential changes in temperature and precipitation. Three scenarios are simulated: an increase of 2°C, an increase of 2°C with a 20 per cent increase in precipitation, and an increase of 2°C with a 20 per cent decrease of precipitation. These changes are well within the range of GCM scenarios for southern Africa by the end of the next century (Hulme, 1996). However, it should be noted that temperature may already differ by several degrees from year to year (Unganai, 1996).

The results show that a temperature increase would have most effect on malaria transmission potential in the areas at high altitudes (> 900 m). In the relatively drier, lower altitudes, a temperature increase of 2°C may result in a decrease in moisture in such a way that parts may become too dry for transmission to take place. This effect is of course more pronounced if the temperature increase is accompanied by a precipitation decrease. The result may be a contraction of the transmission season. In the highlands, transmission increase occurs with all the scenarios.

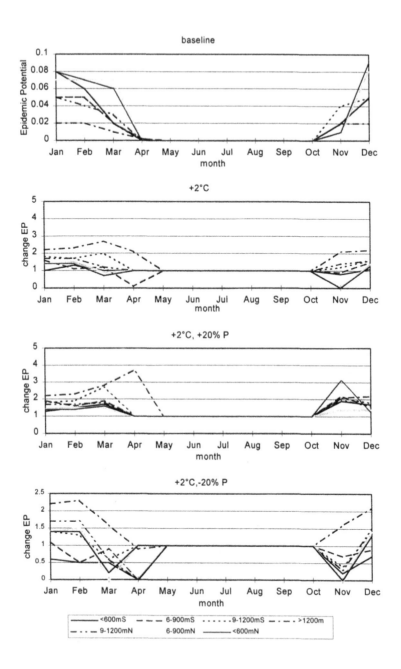

Figure 3.15: Monthly Malarial EP by Altitude Zone under Current Climate and under Climate Change Scenarios

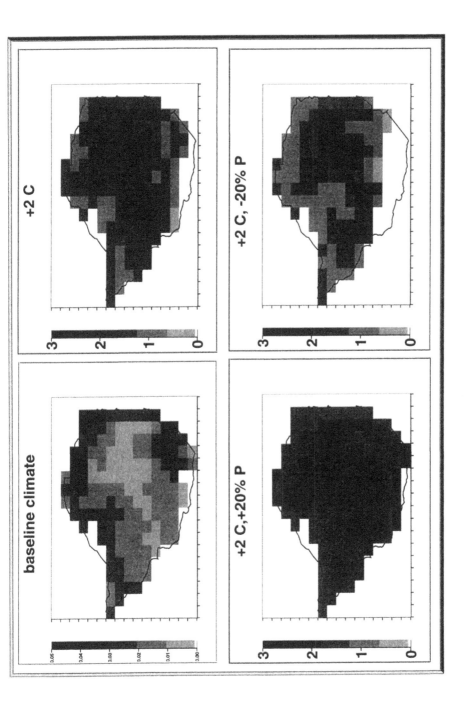

Figure 3.16: Yearly Average Malarial EP for Zimbabwe under Current Climate and under Climate Change Scenarios, Calculated from Monthly Temperatures and Precipitation

71

To examine the impact on malaria incidence, the average incidence data for the years 1972–1981 for the various altitudes from Taylor & Mutambu (1986) are used as the 'equilibrium incidence' for the baseline climate, from which equation 3.19 is calibrated. The level of immunity in the population, as simulated with the model, remains very low in the region of low transmission, which has been confirmed by observation (Freeman, 1995). Using the same climatic changes as above, this equilibrium incidence changes as depicted in Figure 3.17. Again, the yearly incidence of the higher altitudes is most sensitive to changes in climate. Not surprisingly, the drier regions in the south are more sensitive to changes in precipitation than the wetter part of Zimbabwe, which is more centrally located (Unganai, 1996); in the south, in some areas a decrease in precipitation may decrease incidence rates.

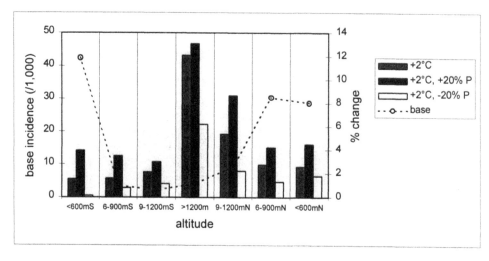

Figure 3.17: Average Annual Malaria Incidence by Altitude Zone for the 1972–1981 (Base) (Taylor & Mutambu, 1986) and under Climate Change Scenarios

Dengue in Selected Cities

There are several sites for which there is some information on dengue transmission. Thailand is an example where cases of dengue are reported from all provinces throughout the year. The transmission is seasonal, with the peak number of cases country-wide falling in July and August (Gunakasem *et al.*, 1981). Using the average temperature data (the baseline temperature data in this section are from 1961–1990 and are extracted from the National Oceanic and Atmospheric Administration (NOAA) baseline climatological data set (Baker *et al.*, 1994)) for centrally located Bangkok as representative of Thailand in the EP model shows transmission, as observed, during the entire year. Furthermore, a strong seasonal pattern is observed too (Figure 3.18). Adding the estimated duration of the growth phase of 3 months to the projected peak of transmission (epidemics are often not recognised until the prevalence of infection rises to perhaps 1 per cent of the population), results in the observed peak number of cases in July and August. Sheppard *et al.* (1969) reported that most cases of DHF/DSS in Bangkok occurred during a 6 month period between May and October. After the growth phase is added to the broad peak of potential transmission (March–July), model projections are consistent with this observation.

Transmission also occurs throughout the year in Puerto Rico with some 55 per cent of all cases occurring in the months of September to December (Anonymous, 1992). While rainfall in southern locations is related to mosquito breeding *(A. aegypti)*, in most areas of Puerto Rico rainfall is adequate, so that the number of water-filled artificial containers is largely independent of precipitation (Moore *et al.*, 1978). The results presented in Figure 3.18, using San Juan as representative of Puerto Rico, are similar to those reported. Climatic changes, as projected by the three GCMs, may result in a longer period of intensive transmission (Figure 3.18) for Bangkok as well as for San Juan.

The Thailand and Puerto Rico examples demonstrate that the EP model can correctly project the endemic maintenance of the virus throughout the year and the existence, timing and duration of the seasonal peaks in transmission. Given the expected latitudinal and altitudinal expansion, the ability of the model to simulate transmission in temperate locations is examined below.

Mexico City, although surrounded at lower elevations by endemic dengue areas, has historically been free of dengue transmission by virtue of the city's altitude (2,485 m) and the related cool climate (Herrera-Basto *et al.*, 1992). Figure 3.18 shows the low EP during the entire year. Although the temperatures in Mexico City are not too cold for mosquito breeding, they are sufficiently low to make the EIP very long. Elevated temperatures may well introduce dengue transmission into this metropolis; in this context the large increase of the EP around April is of particular importance, as in this period transmission potential is already higher than at other times of the year.

Two cities which have experienced epidemic outbreaks of dengue are Athens and Philadelphia. A widespread dengue epidemic of some 650,000 cases took place in Athens in August and September of the year 1928 (Halstead & Papaevangelou, 1980). Two main reasons for this explosive epidemic were the high vector densities due to a severe piped-water shortage at that time, leading to widespread storage of water (providing suitable breeding places for the vector), and the immigration of some 1.5 million repatriated Greeks following the Greco-Turkish war of 1922, introducing the virus (Halstead & Papaevangelou, 1980). In

Philadelphia, an epidemic occurred in 1780, probably as a result of the unusually hot summer during that year and a ship-borne introduction of the virus (Chandler, 1945). Consistent with these observations, model projections suggest that intense transmission could occur in these cities, but only in late summer. Here, the growth phase lasts for about 1 month. An elevation of the EP for both cities is to be expected as climate changes (Figure 3.18). Two climate change scenarios simulate potential transmission in several months in which transmission is not possible under current climatic conditions. However, more important is the increase in transmission potential in the 'high-risk months'. The same analysis, described above, performed using more refined climatic data (Jetten & Focks, 1997), yields similar results.

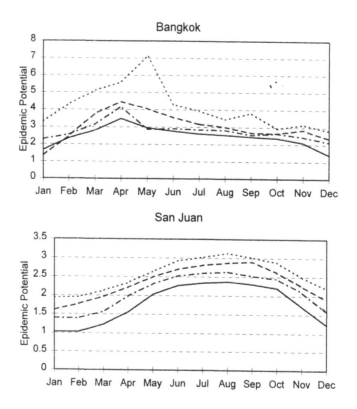

Figure 3.18: Monthly EP for Selected Cities under Current Climate and under Climate Change Scenarios

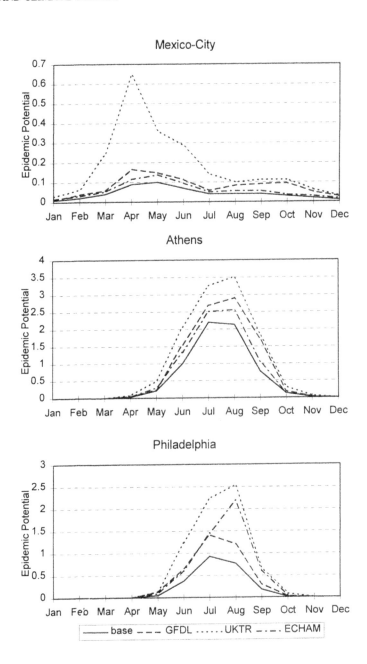

Figure 3.18: *Continued.*

MODEL LIMITATIONS AND UNCERTAINTIES

Modelling the impact of climate change on the complex transmission cycles of vector-borne diseases is beset with numerous uncertainties and several assumptions need to be considered. An important assumption is that all mosquito, snail and parasite species have similar characteristics. This can be justified by the fact that in general the shape of the curve describing the relation between temperature, mortality and infection rates and parasite development is similar for different vector species. However, since dengue involves only one primary vector species, *A. aegypti* (*A. albopictus* being a competent though less significant carrier at present), the global maps of dengue EP may be more generalisable than the assessment of malaria and schistosomiasis risk, since several anopheline mosquitoes and schistosomiasis snail species may transmit the disease.

A number of effects of temperature on vector and human populations are not taken into account. For instance, changes in snail longevity and reproduction rates will influence the age structure of the snail population, which will in turn influence the mean mortality and the mean physical size of the snail population, as well as snail numbers, which thus implies changes in transmission rates of the parasites. Furthermore, climate changes may affect human behaviour as well (reduced water contact rates during the cool season), but changes in human behaviour resulting from temperature increases are not considered here. Neither does the model take account of the indirect effect of climate changes on transmission dynamics, such as changes in irrigation practices and desertification. Since the prevalence of vector-borne diseases, especially malaria and schistosomiasis, in many areas is closely linked to irrigated agriculture, projected increase in rice cultivation in new areas may lead to a further increase in these diseases (Hunter *et al.*, 1993). The displacement of human populations associated with global warming and rising sea levels, and the effect of the evolution of greater resistance to antimalarial drugs and pesticides due to the accelerated life cycles of parasites and vectors at higher temperatures, may further increase the efforts necessary to control these diseases (see also Chapter 4).

Furthermore, the sensitivity of vector-borne disease to changes in climate depends on preceding and coexistent circumstances, such as socio-economic development, local environmental conditions, human behaviour and immunity, and the effectiveness of control measures. For schistosomiasis, even more than for malaria and dengue, such local and specific changes are likely to mask the impact of global climate changes. For example, it is estimated that a 73 per cent reduction in morbidity due to schistosomiasis could be achieved by improving water quality and the availability of sanitation services (WHO, 1993a). This could be reflected by a decrease of the term c_2 in equation 3.2, leading to a proportional decrease in potential transmission.

In interpreting the results, one should note the inability GCMs accurately to simulate precipitation and changes in climate variability, such as changes in the frequencies of droughts, which also could have a significant effect on vector-borne disease transmission, and the omission of sulphate-forcing in the models used. It should be stressed that the temperature scenarios used in this study are not intended to embrace the range of uncertainties attributable to different climate sensitivities, to alternative greenhouse gas emissions or to less tangible sources of error, but rather are illustrative of the differences in geographical and seasonal patterns of climate change (IPCC, 1990). A further simplification is the use of mean monthly temperatures, although daily temperatures may fluctuate by several degrees. However, analyses which use more refined climatological data (daily minimum and maximum temperature) yield similar results (Jetten *et al.*, 1996; Jetten & Focks, 1997).

A final important aspect of modelling climate change impacts on vector-borne diseases is the cumulation of uncertainties in the cause-effect chain: uncertainties in the outcomes of climate change models influence the uncertainties in assessing a climate-related change of the transmission potential of a vector population; uncertainties in both the climate and vector models influence the uncertainties surrounding the estimates of disease prevalence (see also Chapter 2). Figure 3.19 illustrates the cumulation of uncertainties associated with climate change projections, malaria mosquito transmission dynamics, and malaria prevalence, based on variations of only some crucial parameters. This example, for *P. falciparum* in regions of low endemicity, shows how the uncertainty range widens as one moves to more remote parts of the cause-effect chain. The change in malaria prevalence in the year 2050 may range between 1.8 and 2.5 times the 1990 level, based on a variation of only two of the model parameters in the human systems, whereas it may vary between 1.4 and 4.8 times the 1990 level if the uncertainties in the climate and mosquito systems are also taken into account. For *P. falciparum* in regions of high endemicity the total uncertainty in the change of the prevalence varies between 0.9–1.5; for *P. vivax* in regions of low endemicity the range is 1.4–3.5 times the 1990 level (duration of immunity between 1–2 years, duration infection between 2–5 years; survival probability and minimum temperature for parasite development as in Figure 3.6). Despite the large uncertainties, the general trend still shows an increase in mosquito transmission dynamics and prevalence. Only in the simulation of *P. falciparum* malaria in regions of high endemicity, may an increase in immune period (average of 2 years instead of 1.5 years), combined with a decrease in the time being infected (average of 0.75 years instead of 1 year), decrease prevalence rates compared to the initial situation.

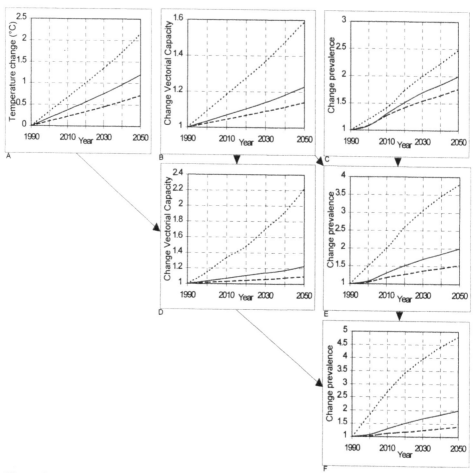

Figure 3.19: Cumulated Uncertainties in Modelling the Impacts of Global Climate Change on Malaria Transmission. (A): The Global Mean Temperature Increase (with Respect to 1990) Using the IS92a Scenario and the Equilibrium Climate Sensitivities of 1.5°C (Low Estimate), 2.5°C (Central), and 4.5°C (High) (IPCC, 1990); (B): Uncertainty Range, for the Central Climate Change Estimate, in Mosquito Transmission Potential, Varying p between 0.8 and 0.95, and $T_{min,m}$ between 16 and 19°C (See Figure 3.6); (C): Uncertainty Range for the Estimated Change in *P.falciparum* Prevalence (per Thousand) in Regions of Low Endemicity, Varying ν between 1 and 2 Years, and τ between 0.75 and 2 Years (see Section on Malaria Prevalence); (D): Cumulated Uncertainty Range of the Vectorial Capacity Change, Taking into Account the Uncertainties of Both the Climate (A) and Mosquito (B) Simulations; (E): Cumulated Uncertainty Range of the Change in Prevalence, Taking into Account the Uncertainties of Both the Mosquito (B) and Prevalence (C) Simulations; (F): Cumulated Uncertainty Range of the Change in Malaria Prevalence, Taking into Account the Uncertainties of the Climate (A), the Mosquito (B), and the Prevalence (C) Simulations

DISCUSSION AND CONCLUSIONS

The extent of climate change impacts on the distribution of vector-borne diseases depends on the climate scenario and specific characteristics of the vector-borne disease concerned. However, all scenarios were found to increase the populations at risk of malaria, dengue and schistosomiasis. Model simulations show an increase from around 60 per cent to around 70 per cent in the proportion of the population in the developing world living within the *potential* malaria transmission zone by the middle of the next century. The climate-related increase in malaria prevalence is estimated to be in the order of 220 million–480 million cases in the year 2050, relative to an assumed global background total of around 800 million–1200 million. The additional number of people at risk of malaria, schistosomiasis and dengue may increase up to ~720 million, ~40 million and ~195 million.

Although it is probably correct to assume that an extension of the areas conducive to vector-borne disease transmission, now particularly confined to the (sub)tropics, will result in encroachment into previously more temperate regions if global warming occurs, it is less certain that the diseases will eventually be as prevalent in the newly invaded areas as they have been elsewhere. However, there is a significant risk of local (re-)introduction of malaria and dengue transmission in developed countries, including large parts of Australia, the USA, and southern and central Europe, which is associated with imported cases of malaria and dengue infected humans, since the former breeding sites of several *Anopheles* and *Aedes* species are still available. The risk of the introduction of the parasite *P. falciparum* in the Mediterranean countries of Europe seems negligible, as anophelines in this region appear to be refractory to it (de Zulueta *et al.*, 1975). The risk of renewed transmission of schistosomiasis in countries in which, until recently, loci of transmission were present (e.g. Japan and Portugal) will increase. Given the fact that effective control measures are economically feasible in these countries, it is not to be expected that human-induced climate changes would lead to a return of a state of endemicity in these areas. Increased vigilance in previously endemic, but not vector-free, areas will be necessary, however.

In view of their high potential receptivity and the immunological naivety of the population, the highest risks of the intensifying of transmission of malaria, dengue and schistosomiasis reside in regions of hitherto no or low endemicity on the altitudinal and latitudinal fringes of disease transmission, as illustrated by the examples of Zimbabwe and the selected 'dengue cities'. Newly affected populations would initially experience higher case fatality rates due to the lack of naturally acquired immunity. Latitudinal fringes include regions in Africa, South-East Asia, and South America. Of particular importance is the increase in EP at higher altitudes surrounded by endemic areas, such as the Eastern Highlands of Africa, the Andes region in South America, and the western mountainous region of China, where an increase in temperature of a few degrees may raise EP sufficiently to transform formerly non-endemic areas into areas subject to seasonal epidemics. Observations of recent malaria emergence, attributed or partially attributed to climatic factors, in Rwanda, Zambia, Swaziland, Ethiopia, Madagascar and Pakistan (Bouma *et al.*, 1994; Loevingsohn, 1994; de Zulueta, 1994), the spread of dengue to recently unaffected higher altitudes in Mexico (Herrera-Basto *et al.*, 1992), and the extension of schistosomiasis in Africa into the previously pristine Lake Malawi (Centers for Disease Control (CDC), 1993) support this conclusion. In the current highly endemic areas, the prevalence of infection is persistently high, and will be only little affected by climate-induced changes in the factors of transmission. Given that there are often insufficient resources to take the adaptive and preventive measures which are required to deal with these diseases adequately, the potential effects of anthropogenic climate change must be taken seriously, especially where the lethal *P. falciparum* parasite is involved.

Chapter 4

MODELLING MALARIA AS A COMPLEX ADAPTIVE SYSTEM

INTRODUCTION

As described in the previous chapter, malaria is one of the world's most important vector-borne diseases, and there are few infectious diseases which have as great an impact on the social and economic development of societies. At present, the distribution of malaria is mainly restricted to the tropics and subtropics, although malaria was a relatively common disease in many temperate areas of the world before World War II.

Although the effective use of DDT and other insecticides after 1945 led to a significant global decrease in the prevalence of malaria, and to its eradication or near-eradication in temperate zones and in some tropical areas, the rate of decrease has now slowed down considerably and a resurgence of malaria has occurred in several countries. The development of resistance to insecticides is considered to be one of the main obstacles in using insecticides for vector control in any strategy of malaria control/eradication. Resistance to insecticides is most pronounced in regions of Africa, Central America, western and South-East Asia (Pant, 1988; WHO, 1992).

A further obstacle is the development of resistance to antimalarial drugs in *P. falciparum*, the malaria parasite responsible for most deaths. For many centuries malaria has been treated with an extract from the bark of the cinchona tree, namely quinine, while a new (synthetic) drug, chloroquine, which became available at the end of World War II, was found to be capable of preventing and curing malaria, especially since it was less toxic and effective in less frequent doses. By the 1960s, however, plasmodia resistant to chloroquine had emerged, and *P. falciparum* which are resistant to the drug are currently found throughout extensive regions of Africa, South-East Asia and South America. The increased selection and progressive dispersal of parasites that are resistant to antimalarial drugs are mainly caused by the fact that these preparations are increasingly being used as prophylactics and for self-medication, usually in insufficient doses. The problem of drug resistance has become particularly alarming in Africa, and its continual exacerbation hampers efforts to provide adequate treatment of the disease (Nájera *et al.*, 1992).

It is evident that malaria patterns have hitherto depended to a large extent on the effectiveness of control efforts, together with socio-economic development. Although new drugs are being developed and work is progressing on various potential malaria vaccines, given the increasing resistance of the malaria mosquito to insecticides on the one hand and of its parasite to antimalarial drugs on the other, the treatment of malaria seems likely to be more problematic in the future. Most of the mathematical models that estimate the effect of eradication programmes on malaria transmission have not yet included the ability of organisms to develop resistance to drugs or insecticides (e.g. Nájera, 1974; Molineaux & Gramiccia, 1980; Collett & Lye, 1987), although the evidence gathered in the field proves that this is a serious omission. However, with respect to resistance dynamics, a number of simulation models have contributed to our general understanding of this phenomenon and the development of strategies to reduce the development of resistance (a review of these can be found in Glass *et al.*, 1984). As the previous chapter has revealed, another factor which may influence future malaria trends is the projected human-induced climate change.

While the model designed to enhance quantitative projections of climate-related changes in the potential distribution of malaria, described in the previous chapter, does take account of how climate change affects the mosquito population – that is, feeding frequency, longevity of the mosquito, and the climatic effect on the incubation period of the malarial parasite inside the mosquito – it does not address artificial interventions by humans and how these may affect the increased malaria risk associated with climate changes. In order to allow for both antimalarial control measures *and* the adaptation of mosquitoes and parasites to such malaria control policies, the simulation model discussed in Chapter 3 is combined with the use of genetic algorithms (Janssen & Martens, 1996, 1997). The latter involves a general and robust evolutionary modelling approach which is based on the mechanics of the survival of the fittest, whereby the inclusion of the notion of variability within the population renders the genetic algorithm a suitable tool for simulating the adaptive behaviour of a population within a changing environment. This chapter presents a simplistic, idealised model of the resistance cycles associated with insecticide and drug use in malaria control programmes, together with the impact of climate changes. Although this approach is adopted solely for heuristic purposes, it nevertheless does succeed in elucidating the mechanism of resistance development, interactions associated with climate change, and consequences for the implementation of strategies in malaria management.

THE GENETIC ALGORITHM

Because insecticides and antimalarial drugs are agents of selection, resistance to them can be studied using the same theoretical frameworks as those which have been applied to other types of evolutionary change. Consider those organisms which evolve by means of two primary processes, namely, natural selection and sexual reproduction. The first process determines which members of a population survive to reproduce, while the second ensures mixing and recombination among the genes carried by the parents' offspring. An obvious biological consequence of sexual reproduction is the generation of new combinations by mixing genetic information from different individuals. Without this mixing, adaptive evolution would simply consist of the sequential selection of mutations in the genetic information. Selection takes place on the basis of the fitness of the organisms.

Holland (1975) has tried to abstract and explain the adaptive processes of natural systems and developed a mathematical representation of these processes: the genetic algorithm. The basic construction is to consider a population of individuals in which each individual represents a potential solution to a problem. The relative success of each individual vis-à-vis that problem is considered to be an indicator of that individual's fitness, and, by virtue of natural selection, the most fit individuals produce similar, but non-identical, offspring for the next generation (Hofbauer & Sigmund, 1988). This concept of the survival of the fittest, as simulated by the genetic algorithm, is the main reason why genetic algorithms are chosen to model the adaptation

According to Goldberg (1989), genetic algorithms may be successful robust algorithms in optimisation because they are able both to select strings with useful blocks of information, and to concentrate their search (selection) on variations which include those blocks. The genetic algorithms test and exploit large numbers of regions in the search space while manipulating relatively few strings, and without requiring specific information about the functional forms of the search space. Instead of using the genetic algorithm purely as an optimisation routine, its power as an optimiser is used to simulate processes within a changing environment, more specifically to simulate the basic elements of evolutionary processes of resistance development of mosquitoes and parasites in a changing environment.

Box 4.1:
The Working of a Genetic Algorithm

The following illustrates the working of a genetic algorithm (Figure 4.1). Consider, for example, a population of N mosquitoes, each represented by a chromosomal string of L allele values (= number of values of which the gene is composed). An initial population is constructed at random on a specific range: generation g_0. The genetic algorithm then performs two operations. First, its selection algorithm uses the population's N fitness measures to determine how many offspring each member of g_0 contributes to g_1. Second, a set of genetic operators is applied to the offspring to make them different from their parents. The resulting population is now g_1. These individuals are again evaluated in the next time step according to the new situation, and the cycle repeats itself.

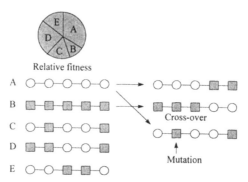

Figure 4.1: Schematic Diagram of a Genetic Algorithm (Janssen, 1996)

The genetic algorithm can be formulated in a more formal way:

(1)   An individual can be characterised by a bit string of fixed length L, which is denoted as a, and $a \in B^L$ where $B = \{0,1\}$. The bit string can be separated into n segments of equal length l, thus implying that $L = n \times l$. Each segment i is interpreted as the binary code of the object variable $x_i \in [u_i, v_i]$ which can be decoded by applying:

$$\Gamma_i(a_{i1}...a_{il}) = u_i + \frac{v_i - u_i}{2^l - 1} \cdot \left(\sum_{j=0}^{l-1} a_{i(l-j)} 2^j \right) \qquad (4.1)$$

where $(a_{i1}...a_{il})$ denotes the i-th segment of an individual $a \in B^L$. Then $\Gamma = \Gamma_1..\Gamma_n$ yields a vector of real values on the desired range $[u_i, v_i]$.

Example: $a_i = 10011$, $u_i = 0$, $v_i = 1$
            $\Gamma = (1 \times 2^4 + 0 \times 2^3 + 0 \times 2^2 + 1 \times 2^1 + 1 \times 2^0)/31 = 0.613$

(2)   Mutation is a bit reversal event that occurs with the small probability of p_m per bit. This mutation can explore new genetic information and is a powerful operator in discovering ways to adapt to a changing environment.

Example: Suppose we have the following binary bit string: 11111
            At random, roughly one in every 1000 symbols
            flips from 0 to 1 or vice versa ($p_m = 0.001$)
            in our example from 1 to 0:   11011

(3)   The algorithm uses a cross-over operator that exchanges substrings arbitrarily between two individuals with a probability p_c. Both the length and position of these substrings are chosen at random, but are identical for both individuals.

Example: Suppose we have the following bit strings: 11111 and 00000
            A point along the strings is selected
            at random and the offspring contains
            a mixture of the parents: 11000 and 00111

Box 4.1 Continued

(4)   The probabilistic selection operator ushers in the next generation by copying individuals on the basis of fitness-proportional probabilities:

$$p_i = \frac{F(a_i)}{\sum_{j=1}^{N} F(a_j)} \qquad (4.2)$$

where $F:B^L \to R$ is the fitness function. The less fit individuals are therefore less likely to reproduce their genetic information.

MODELLING ADAPTATION BY GENETIC ALGORITHMS

In order to incorporate adaptation to antimalarial drugs and insecticides, the dynamic system described in the previous chapter is coupled to genetic algorithms which permit the simulation of the genetic variety within the mosquito population and the parasite population (Figure 4.2). The genetic algorithms determine parameters that in turn determine the resistance of the mosquitoes and parasites and the optimum temperature for mosquito survival. In the experiments discussed in this chapter, attention is restricted to *P. falciparum* since it is the most lethal malaria parasite and is exhibiting world-wide development of resistance to antimalarial drugs. For each subject of adaptation (temperature change, use of insecticides and use of drugs) a genetic algorithm is employed in modelling the transmission of genetic information by means of sexual reproduction.

If genetic algorithms are to be used to simulate the dynamics of malaria, the validity of a number of assumptions must first be considered (see also Taylor (1983), who discussed the lack of experimental data with which one could validate the modelling approach to the issue of resistance development):

- The values adopted for cross-over probability (p_c) and mutation probability (p_m) are imaginary numbers and cannot be validated by empirical research. The selected values are consistent with those generally used in genetic algorithm applications, and the results are not sensitive to these assumptions, as will be shown later in this chapter.
- There is a lack of knowledge about the various shapes of the fitness functions, and those discussed in the following subsections, although mimicking observed patterns (Curtis *et al.*, 1978; Schapira, 1990), are therefore rather subjective and should be regarded as being of illustrative value only.

- The question of which population size is adequate for simulating the variety within a population most be resolved. After testing various numbers, a population of 100 'individuals', each with a length of ten bits, seemed to be appropriate. Of course, the 'real' numbers of mosquitoes or parasites are not simulated by reference to such a group; nevertheless this population size within the genetic algorithm allows the simulation of the aggregate adaptive behaviour of a representative heterogeneous group of mosquitoes and parasites.

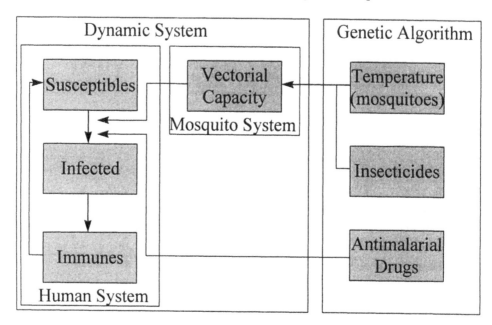

Figure 4.2: Schematic Diagram of the Dynamic Malaria Model Coupled to Genetic Algorithms

The Mosquito Population

The potential of the mosquito population to transmit malaria is now not only determined by temperature, but also by the use of insecticides, $u_1(t)$. The formulation of VC in equation 3.19 is multiplied by the relative fitness of mosquitoes given the application of a certain dose of insecticides, $F^m(u_1)$, resulting in a reformulation of vectorial capacity which includes the impact of the use of insecticides. Furthermore, an adaptive representation of survival probability (p_a) can be used to describe the adaptation of a mosquito population to a change in temperature:

$$VC = k_1 \frac{a^2 p_a^n}{-\ln(p_a)} F^m(u_l)$$

(4.3)

With the help of the genetic algorithm, sexual reproduction is implemented using the two genetic parameters: cross-over probability (p_c) and the mutation probability (p_m). To simulate the adaptation of mosquitoes a crossover probability of 0.4 and a mutation rate of 0.001 were assumed. The two pressures on the mosquito population which are distinguished, namely temperature change and insecticide use, are assumed to be independent of each other.

Adaptation to Insecticides
An important human-induced pressure on the mosquito population is the use of insecticides. Several models have been developed to enable us to understand and manage the evolution of insecticide resistance, and nearly all of them assume that resistance is controlled by two alleles at one locus (Taylor, 1983; Anderson & May, 1991). However, here the fitness function is based on the study published by Tabashnik (1990), who has investigated three- and four-allele models.

Our model is simulated by distinguishing three kinds of mosquitoes, namely susceptible, moderately resistant and resistant individuals, taking them as three classes of individual sensitivity to insecticides. The assumption is that a certain dose of insecticide reduces fitness in the manner depicted in Figure 4.3, whereby it is assumed that the same dose would have a more pronounced impact on susceptible mosquitoes than on (moderately) resistant ones. The fitness function expresses the notion that the fitness of the three classes drops in a decreasing rate for a higher dose of insecticide. Obviously, if alternative insecticides are applied which affect the three categories differently for some reason, being more effective, for example, the results and conclusions may differ.

Table 4.1: Fitness of the Mosquito Population

Class	Relative biotic fitness, $F_{bio}{}^m$	Relative fitness under insecticides, $F_{ins}{}^m$
Susceptible	1.0	$1-u_1/(0.002+u_1)$
Moderately resistant	0.95	$1-u_1/(0.05+u_1)$
Resistant	0.9	$1-u_1/(0.15+u_1)$

In addition, the simulation incorporates a 'biotic fitness' component which represents the relative fitness of the mosquito in the event of no insecticides being used at all. A lower value for the biotic fitness of the more resistant genes explains the lower density of these genes in an insecticide-free environment. Given an initial random distribution, the Table 4.1 is derived for the fitness of mosquitoes, to which a certain dose of insecticides u_1 is applied. It is assumed that 99 per cent of the mosquitoes are susceptible, 0.9 per cent are moderately-resistant and 0.1 per cent resistant in the initial situation (values based on Tabashnik, 1990). The fitness function for a mosquito $F^m(u_1)_i$ is the product of the 'biotic' and 'insecticide' fitness; the average fitness of the individual mosquitoes, $F^m(u_1)$, is used in the equation for VC.

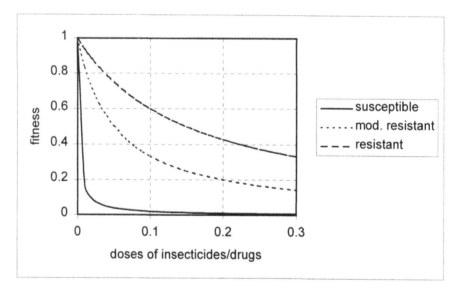

Figure 4.3: Relative Fitness of Mosquitoes Related to the Use of Insecticides: a Certain Dose of Insecticides or Drugs Leads to a Reduction in Fitness which is More Severe in the Case of Susceptible Rather than Resistant Individuals

Adaptation to Temperature Change

For every mosquito there is a temperature at which its expected lifetime (i.e. survival probability) would be maximised (Figure 3.3C), but within the mosquito population there is variation of these optima among individuals. If the temperature increases over a longer period (a few years, say), mosquitoes for which the optimum is higher than average exhibit greater fitness. Due to the mechanisms associated with the 'survival of the fittest', the average optimum temperature for longevity will therefore rise. (This account of adaptation to unfavourable temperatures is just one of the possibilities. Another would be the migration of mosquitoes to microhabitats where temperatures are more suitable.) The implementation of this process by means of a genetic algorithm proceeds as follows.

Within the mosquito population, the daily probability of survival is a function of temperature (see equation 3.6). Within the population, individual temperature optima are scattered around the mean temperature. For simplicity, no distinction is made between seasonal temperature changes. The daily survival probability can therefore be treated as a function of the *local mean temperature*. This results in a daily survival probability such that the fitness function of mosquito i becomes,

$$F_{T,i} = -4.4 + 1.31(T - T_i^a) - 0.03(T - T_i^a)^2 \tag{4.4}$$

The variable T_i^a represents the individual adaptation to temperature. If temperature T changes, the value of T_i^a will also change since the 'survival of the fittest' keeps the mosquitoes in the optimum temperature zone. Then, the daily survival probability of the adult mosquito becomes:

$$P_a = \exp\left(\frac{-1}{-4.4 + 1.31(T - T^a) - 0.03(T - T^a)^2} \right) \tag{4.5}$$

where T^a is the mean of T_i^a.

The Parasites

In the model described in the previous chapter (Box 3.3), the rate at which individuals become infected depends on the VC, which represents the transmission potential of the mosquito population, and the proportion of infected people in the human population (see equation 3.16). In the complex adaptive systems approach, the infection rate also depends on the amount of drug use and the sensitivity of malarial parasites to such drugs (i.e. the fitness of the parasites). $FP(u_2)$ represents the fitness, which may decrease in the event of antimalarial drug use, depending on the degree of resistance. The use of drugs thus leads to a decrease in the infection rate and consequently an increase in the rate of loss of immunity and in the rate of loss of infection (equations 3.17 and 3.18).

$$\lambda(t) = VC(t)Y'(t)F^p(u_2) \tag{4.6}$$

The dynamics of the gene pool in parasites differ from those in the mosquitoes. Since the population of parasites is spread among the human population and the mosquito population, the transmission of resistant parasites through a vector population to other human hosts limits the efficacy of adaptation in the parasite population at large. Note that a single gene pool for parasites is used, although several local clusters do exist (in the hosts). In view of the lack of data, the same cross-over and mutation probabilities are used as for the mosquitoes in the reference.

Adaptation to Drugs

The adaptation of parasites to the use of antimalarial drugs is modelled in a similar manner as the modelling of the resistance among mosquitoes to the use of insecticides. Thus, a three-phenotype model is simulated by distinguishing three kinds of parasites, namely susceptible, moderately resistant and resistant individuals, and taking these as three classes of individual sensitivity to the drugs involved. Given an initial random distribution, Table 4.2 gives the fitness of parasites to which a certain dose of drugs u_2 is applied, assuming that 99 per cent of the population is susceptible, 0.9 per cent is moderately-resistant, and 0.1 per cent resistant in the initial situation.

Table 4.2: Fitness of the Parasite Population

Class	Relative biotic fitness, $F_{bio}{}^p$	Relative fitness under insecticides, $F_{ins}{}^p$
Susceptible	1.0	$1-u_2/(0.002+u_2)$
Moderately-resistant	0.95	$1-u_2/(0.05+u_2)$
Resistant	0.9	$1-u_2/(0.15+u_2)$

Similarly, the fitness function for a parasite $Fp(u_2)_i$ is the product of the 'biotic' and 'drugs' fitness; the average fitness of the individual parasites, $Fp(u_2)$, is used to determine the impact of resistance on the transmission dynamics within the human population. It should be noted, however, that in some places the biological advantage of chloroquine-resistant *P. falciparum* has been observed (discussed by Wernsdorfer, 1994). This would imply that resistance development would proceed more rapidly than under the assumptions discussed above.

Migration and Refugees among Mosquitoes and Parasites

Georghiou and Taylor (1977) argue that the migration of insects tends to delay the rate of evolution of resistance. In addition, the percentage of mosquitoes or parasites not reached by the antimalarial treatment (the so-called refugees) will inevitably influence resistance development. The complex adaptive systems approach takes account of both of these processes in the development of resistance, among mosquitoes as well as among parasites.

It would seem self-evident that, depending on landscape and infrastructure, mosquitoes are more or less able to migrate from place to place, and that mosquitoes susceptible to insecticides may thus enter a treated area. Moreover parasites susceptible to antimalarial drugs can also migrate, whether they are carried by mosquitoes or humans. Migration is modelled by assuming that during each time step a fraction of the new population is bred under the initial conditions, i.e. not yet adapted to the changed conditions. Of course, migration may also introduce resistant parasites/mosquitoes in areas where they were previously absent.

Insecticides are sprayed on specific areas so that 100 per cent coverage is seldom achieved. Drugs are not taken (sufficiently) by all humans so that a fraction of the parasites escapes the drug effect. This phenomenon of refugees is modelled by assuming that during each time step a part of the population, the size of which is randomly selected, has not been treated, despite the control programmes which have been implemented.

MODELLING EXPERIMENTS

Introduction

The experiments deal with the consequences of the use of insecticides and anti-malarial drugs, together with temperature change, for the occurrence of malaria for a time horizon of one decade, using time steps of 0.1 years. Although malaria situations are extremely heterogeneous with respect to resistance to change, the two types of regions distinguished are a region of low endemicity and a region of high endemicity. Although the real generational longevity among the parasites and mosquitoes is not specified, the time horizon is based on the observed time lapse in acquiring resistance (see Table 4.3). The other parameter values are described in Chapter 3. (In the experiments described below, only two age classes are distinguished: 0–5 and above 5 years.) The simulated areas of lower endemicity can be characterised as exhibiting low VC, resulting in a high percentage of susceptible persons (~80 per cent), and low percentages of infected (~8 per cent) and immune persons (~12 per cent). Areas of low endemicity vis-à-vis *P. falciparum* can be found in South-East Asia and South America. Regions of high endemicity are characterised by a relatively high VC. In the initial situation there is a high percentage of immune (~68 per cent) and infected persons (~27 per cent). The younger age class especially suffers from a high percentage of infection (~45 per cent). Highly endemic regions are mainly found in tropical Africa. In the starting year, the situation is assumed to be near equilibrium. This assumption of an equilibrium state is made for analytical purposes, namely, to render the impact of control policies and temperature change on the occurrence of malaria transparent, thereby including the adaptation of mosquitoes and parasites. Therefore, a steady-state situation in demographic, social and economic development is assumed, although these factors may influence future developments of malaria.

The results are presented as time series covering a period of 10 years. In view of the stochastic elements of the model a large number of runs (100) are used, and the mean and the extremes of important indicators are determined. (Experiments showed that a higher number of runs would not affect the mean values significantly.) This procedure yields ranges of uncertainty, whereby the uncertainty does not lie in the different parameter values of the model, but rather in the stochastic characteristics and the complexity of the system.

Table 4.3: The Time that Elapses before a Majority (i.e. > 50 Per Cent) of the Individuals in the Mosquito Population Become Resistant to the Control Agent (Brown & Pal (1971))

Anopheline mosquitoes (different localities)	Control agent	Time to resistance (years)
A. sacharovi	DDT	4–6
	Dieldrin	8
A. maculipennis	DDT	5
A. stephansi	DDT	7
	Dieldrin	5
A. culicifacies	DDT	8–12
A. annuaris	DDT	3-4
A. sundaicus	DDT	3
	Dieldrin	1–3
A. quadrimaculatus	DDT	2–7
	Dieldrin	2–7
A. pseudopunctipennis	DDT	>20
	Dieldrin	18 weeks

In the interest of analytical lucidity two broad hypothetical control levels for both insecticides and antimalarial drugs are considered, namely the low and the high dose. In case of a low dose, u_i is equal to 0.002, which represents a 50 per cent deterioration in the fitness of susceptible mosquitoes or parasites. The high dose, u_i, is assumed to be equal to 0.05 such that the fitness of the moderately resistant mosquitoes or parasites decreases by 50 per cent. A typical outcome is given in Figure 4.4, which shows the impact of using a low dose of insecticides. Although the input variables are the same for the 100 runs, there is a large spread in the optimum temperature for the mosquitoes, the vectorial capacity and the incidence of malaria. Although on average the use of a low dose of insecticides leads to an increase in the incidence of malaria in the long run, it might also lead to a slow decrease of the incidence if evolutionary adaptation among mosquitoes proceeds very slowly. In order to envisage the trends for the various sensitivity tests only the average scores are depicted in the following subsections.

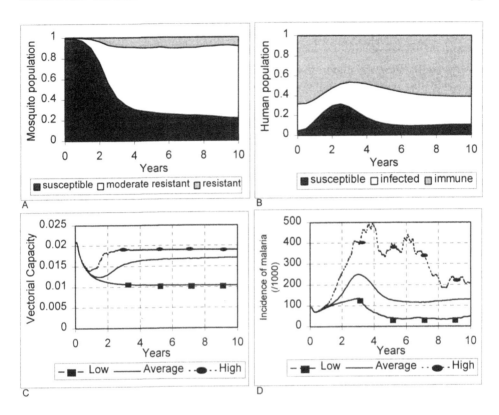

Figure 4.4: An Example of an Experiment in a Region of High Endemicity: Depicted Are the Average Fractions of Resistant, Moderately Resistant and Susceptible Mosquitoes (A), and the Average Fractions of Immune, Infected and Susceptible People (B), for a Sample of 100 Runs; Furthermore, the Average and Extremes Found for the Change in VC (C) and Malaria Incidence (D) Are Shown, also for a Sample of 100 Runs

Sensitivity to Migration

Migration of mosquitoes and parasites can influence the development of resistance. Comins (1977), for example, showed that the migration of insects may greatly retard the development of insecticide resistance, and recent observations in Papua New Guinea and Tanzania support such model-based hypotheses (Paul *et al.*, 1995). Various studies (e.g. Comins, 1977; Tabashnik & Croft, 1982) found two distinct phases in the time required to develop resistance. At low doses resistance developed more rapidly as the dose increased, paralleling the case in which migration is absent; this in contrast to the case of high doses, in which resistance develops more slowly as the dose increases. In the absence of migration, the rate of resistance development is determined primarily by the rate at which susceptible genes are removed from the population. As the dose increases, susceptible genes are removed more rapidly, and resistance consequently develops apace. The pattern is similar at low doses in the presence of migration. Where migration is present and doses are high enough to kill heterozygotes (which are intermediate between the susceptible and resistant genes, comparable with 'moderately resistant' in this chapter), however, the high dose also removes resistant genes from the population. As dose increases in this range, more heterozygotes are killed, leaving relatively few resistant individuals. The influx of susceptible immigrants retards the further development of resistance.

The impact of mosquito migration on the development of insecticide resistance is analysed by postulating various levels of insecticide application and various percentages of migration and subsequently calculating the number of time steps required for 50 per cent of the genes to achieve resistance. The results are depicted in Figure 4.5 and show, as expected, that the migration of susceptible mosquitoes impedes the development of resistance. Furthermore, at high levels of migration (>40 per cent inflow of susceptible mosquitoes) the development of insecticide resistance among the mosquitoes will be entirely blocked. The same pattern is simulated with respect to gene flow (migration) in the parasite population.

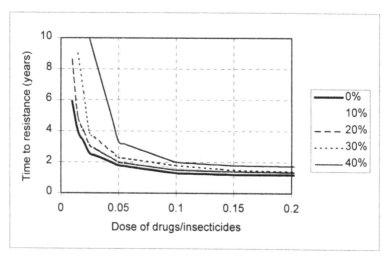

Figure 4.5: Effects of Dose on the Rate of the Evolution of Resistance Featuring Various Percentages of Migrants per Time Step

That the results do not show the two distinct phases, which were found by Comins (1977) and Tabashnik and Croft (1982), is a consequence of the different fitness functions for the various genes. (Comins (1977) and Tabashnik and Croft (1982) do not actually employ the term 'fitness function', but it is considered to be equivalent to their 'dose-mortality lines'.) The relative fitness among the various gene combinations remains rather the same along the line of increasing doses of insecticides. This is not the case where models such as the one adopted by Tabashnik and Croft (1982) are concerned, since heterozygotes are not killed at low doses, but only at high ones. In fact, in such models there is a kind of threshold value in the fitness function (survival rates for the different types of genes), while in the model presented in this chapter a more gradual decrease of the fitness function is assumed.

Sensitivity to the Coverage Rate

In the absence of refugees from control programmes (i.e. 100 per cent coverage), rates of insecticide resistance increase with increasing doses. If, however, a fraction of the mosquito population evades treatment by becoming 'refugees', the development of resistance is expected to be impeded. Tabashnik (1990), for example, shows that if 10 per cent of the mosquitoes are refugees evading exposure to insecticides, this may significantly impede the development of resistance.

The impact of the coverage rate is explored for the different doses applied in various control programmes, and the results are depicted in Figure 4.6. For each time step a certain fraction of the mosquito population is not reached by the control measures, and two distinct phases in the time required to develop resistance are found. In the case of low doses and low percentages of refugees, the results are about the same as in the case of zero refugees. However, when higher doses are applied the time period required to develop resistance lengthens rapidly. The doses of control which mark the two distinct phases are different for each of the various fractions of refugees. Where higher percentages of refugees are concerned, the period of time required to develop resistance starts to become greater at an earlier juncture. Resistance will not develop at all among more than 50 per cent of the refugees. The same pattern results for the malaria parasites. The rate of evolution of resistance by *P. falciparum* could be retarded by selective treatment of those people with high parasitaemias.

An explanation for the existence of these two distinct phases, which are also found by Tabashnik (1990), is the fact that during the period in which the agent (parasite/mosquito) evades treatment, the benefits of being resistant do not hold. In other words, the parasite and mosquito will not benefit from being resistant in periods during which there is no pressure in terms of drugs or insecticides. On the contrary, during such periods, susceptible agents enjoy a higher biotic fitness than resistant ones. By the same token, in the periods during which the parasite/mosquito population is reached by drugs/insecticides, a resistant individual enjoys the benefits of higher fitness. In the case of higher doses, the difference in fitness in the two cases (reached or not reached by a control programme) becomes greater, resulting in an increase in the time required to develop resistance. Furthermore, the presence of a higher fraction of refugees decreases the average time during which the population in general profits from the availability of resistant genes, consequently impeding the development of resistance.

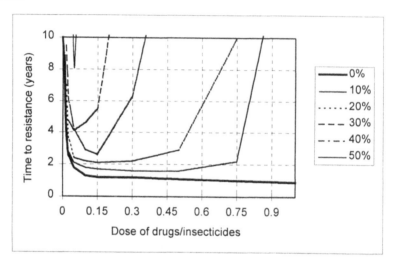

Figure 4.6: Effects of Dose on the Rate of Evolution of Resistance Featuring Various Percentages of Refugees per Time Step

Sensitivity to Temperature Change

For regions of both low and high endemicity the change in the occurrence of malaria is simulated in the event of a temperature increase of 0.25°C or 0.5°C within a decade (Figure 4.7). This temperature increase lies within the range projected by the IPCC (1991). As also shown in Chapter 3, the impact of temperature change on the incidence of malaria would be significant. VC would increase due to increasing biting rates and shorter incubation periods, which would lead to an increase of the incidence of malaria in the order of 50 per cent to 100 per cent in regions of low endemicity. In regions of high endemicity the incidence of malaria would fall by about 15–30 per cent in the event of such temperature changes as a result of the increase in the collective immunity.

The optimum temperature for mosquito survival would increase by 0.05°C in areas of low endemicity and 0.12°C in highly endemic areas, a development which lags behind the local mean temperature increase. The impact of mosquito adaptation to temperature change compared with no adaptation is not significant. This is caused by the fact that only the survival probability p changes as a result of adaptation, and this is only one of the factors (besides the man-biting habit and the

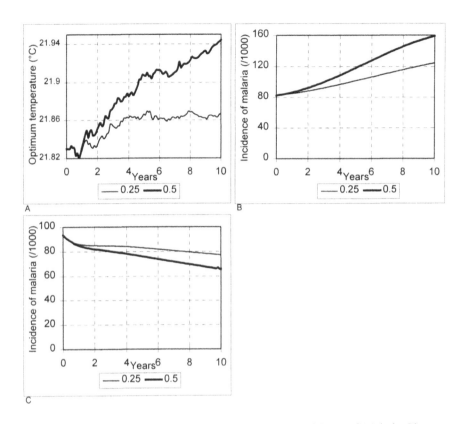

Figure 4.7: Optimum Temperature for Mosquito Survival (A), Incidence of Malaria Given a Region of Low (B) and High (C) Endemicity for a 0.25 and a 0.5 °C Temperature Increase

incubation period) which are related to temperature fluctuations.

An important factor influencing the rate of the evolution of resistance is the number of generations per year: resistance develops faster as the annual number of generations increases. Increasing temperatures will lead to a more rapid succession of generations per unit of time (Jetten & Takken, 1994). As a consequence, resistance development in the malaria mosquito population as presented in this book would have taken place within a shorter time span if a temperature-dependent generation time had been implemented.

Sensitivity to Initial Resistance

The sensitivity of the initial fraction of resistant mosquitoes and parasites was investigated by an illustrative experiment. A low and a high dose of insecticides is postulated in a highly endemic region. Figure 4.8 shows that a higher level of

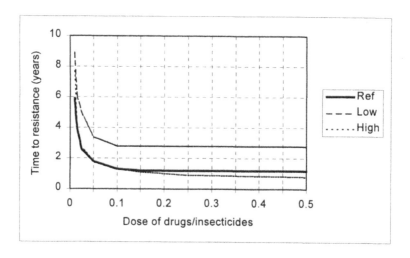

Figure 4.8: Time to Resistance for Different Levels of Initial Resistance, Where *Low* Represents 0.01Per Cent Initial Resistant Genes, *High* Represents 1 Per Cent Initial Resistant Genes, while the *Reference* is 0.1 Per Cent Initial Resistance Genes

initial resistance leads to a more rapid penetration of resistant genes in the mosquito population. Due to a higher selection pressure on the population, the time required to develop resistance decreases in an environment of higher doses. This finding corresponds with those reported in Tabashnik (1990).

Sensitivity to the Operators of the Genetic Algorithm

Within the range of reasonable values of the main operators of the genetic algorithm, a number of values are selected to test the sensitivity of the solutions vis-à-vis the chosen values. As depicted in Figure 4.9A and B, the results do not seem to be sensitive to the chosen values, although some variation does occur. This supports the chosen (unknown) values of the operations.

Implications for Parasite and Vector Control Programmes

If mosquitoes and parasites do not adapt to the use of insecticides and drugs, the new equilibrium can be calculated, given that constant levels of insecticides and/or drugs are used. Because the impact of both control programmes is modelled in a similar manner they have identical effects. The control programme will lower the rate of infection as a result of causing the mosquitoes and/or parasites to be less fit, and because of the decrease in the percentage of infected persons. The percentage of immune persons will likewise decrease, resulting in an increase in the size of the fraction of susceptible humans.

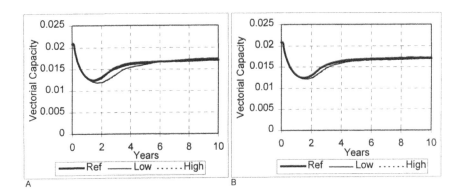

Figure 4.9: VC for Different Values of: A, the Cross-Over Probability, Where *Low* Represents 0.2, *High* 0.8 and *Reference* a Probability of 0.4; and B, the Mutation Probability, Where *Low* Represents 0.0001, *High* 0.01 and *Reference* a Probability of 0.001

The incidence of malaria will decrease in regions of low endemicity as a consequence of the control programmes (Figure 4.10A). In regions of high endemicity an increase of malaria may occur if the control programmes are not stringent enough, the effect being a steeper increase in susceptible humans (immune persons lose their immunity) relative to the decrease in the infection rate (Figure 4.10B).

As a result of the ability of vectors and parasites to adapt to the control programmes, their effectiveness decreases in such a manner that the new equilibria are located nearer to those obtaining in the absence of control programmes. Adaptation may eventually lead to higher incidence rates than those which obtain in the absence of adaptation. A notable exception is the case of low doses in regions of high endemicity since adaptation will then result in a less pronounced increase in the susceptible population, which will exceed the reduced decrease in the infection rate, leading to lower incidence rates.

Figure 4.11 shows the average values over time for different levels of control programmes. In regions of low endemicity the VC first decreases, but due to adaptation among mosquitoes it subsequently increases, albeit to a level that lies somewhat below the initial level. The result is a similar pattern in the incidence of malaria, although the level continues to fall (gradually) as a result of the VC reaching a level below the critical value needed to maintain the infection. It is thus evident that a combination of both drugs and insecticides at low levels is more efficient than high-level use of only one of the two, a finding which reflects the enhanced development of resistance at higher doses.

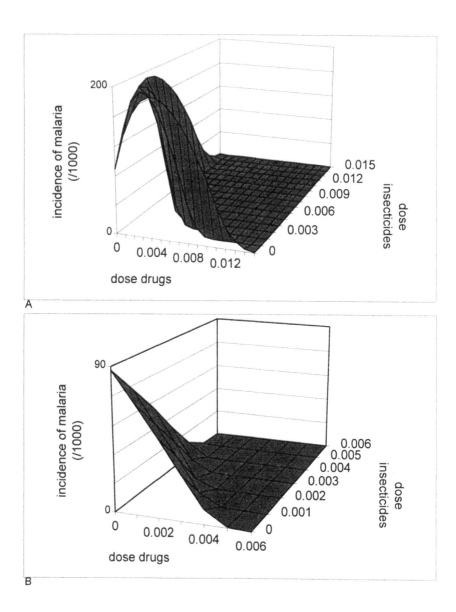

Figure 4.10: Incidence of Malaria for Different Levels of Control in Case of no Adaptation, for Regions of High (A) and Low (B) Endemicity

In regions of high endemicity the decrease in VC exhibits a similar pattern to that which obtains in regions of low endemicity. (One would expect that resistance development would differ in the two regions due to a difference in the gene pool. Nevertheless, for simplicity's sake the same fixed population size is used within the genetic algorithm and therefore the results are similar. An improvement

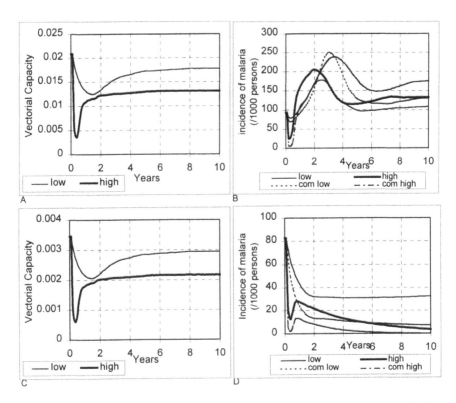

Figure 4.11: VC and Incidence of Malaria for Different Levels of Control in a Region of High (A-B) and Low (C-D) Endemicity; Scenarios *Low* and *High* Depict the Results of a Low or High Dose of Insecticides; for the Scenario *Com Low*, Low Doses of Insecticides are Combined with Low Doses of Drugs; for the Scenario *Com High*, High Doses of Insecticides are Combined with High Doses of Drugs

of the model might be the coupling of VC and the population size of the genetic algorithm.) Due to the difference in the profiles of the populations, the patterns of incidence of malaria are quite dissimilar. Following a reduction in incidence at the outset of the control programmes, incidence subsequently shows an increase due to the lower effectivity of the control measures. Due to the high fraction of susceptible humans after a successful period of control, again as a result of the flow of immune persons due to the increased rate of immunity loss, incidence may even rise to surpass the initial level. In the long run, a combination of two low levels of control does not achieve a better performance than control by a single method. Indeed, incidence peaks at a level even higher than the initial (pre-control) level due to the greater number of susceptible humans who become re-infected.

Adaptive Malaria Management

This subsection analyses the impact of the combined effects of climate change and resistance development among mosquitoes and parasites on the incidence of malaria. This analysis is performed using an adaptive management style, i.e. one which regulates the level of control programmes according to the observed state of the system. Since in the complex adaptive malaria model the resistance development dynamics are implemented in an identical manner for both mosquitoes and parasites, there is a need to consider only one of the two in the analysis: mosquitoes in this case.

The use of insecticides is related to the observed incidence of malaria, and there are two levels of application: a zero dose and a high dose. The assumption is made that if the incidence of malaria fell below 20 per 1000 persons, the use of insecticides would be stopped, while if malaria once again were to exceed this level, it would be re-introduced (at high dose levels). Furthermore, if the incidence of malaria exceeded the level of 100 per 1000 persons, which is above the initial level, the use of insecticides would be stopped as not being effective. The results set out in Figure 4.12 illustrate that in the simulated areas of low endemicity an adequate use of drugs and insecticides may lead to successful control of malaria occurrence. However, if the temperature were to increase by some 0.5°C within a single decade, the efforts to control malaria would have to be intensified significantly. In areas of high endemicity the control of malaria fluctuates during the decade while the incidence continues to fluctuate around the level of 100 per 1000 persons regardless of any temperature increase.

This modelling exercise thus shows that it may not be possible to eradicate malaria in regions of high endemicity using the assumed (i.e. adaptive) management style. However, in regions of low endemicity malaria could be reduced significantly using adaptive management, although increased efforts would be needed in the event of climate change.

DISCUSSION AND CONCLUSIONS

Models can be useful, especially if the opportunity to perform experiments in laboratories or in the field is limited. This is certainly the case where the growing problem of resistance development among malaria vectors as well as malaria parasites to control programmes is concerned, and much remains to be elucidated. Current malaria modelling approaches, however, do not explicitly address the evolutionary character of the development of resistance. The malaria assessment model presented in this chapter is neither comprehensive nor predictive, but rather intended to include evolutionary processes of resistance development in order to provide insights into this complex adaptive system and thus help us to arrive at a

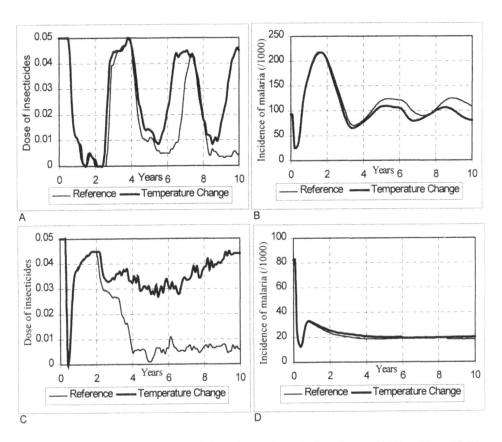

Figure 4.12: Control Patterns and Malaria Incidence for a Region of High (A-B) and Low (C-D) Endemicity in the Event an Adaptive Management Style is Being Adopted; the Two Different Lines Show the Impact of a Projected Climate Change on the Adaptive Control Programmes

better understanding of the possible effects of control programmes.

The analysis distinguishes between two exemplary malaria regions, although malaria situations are extremely heterogeneous with respect to resistance to change. The results for the two situations described in this chapter suggest that *adequate* use of insecticides and drugs may reduce the occurrence of malaria in regions of low endemicity, although increased efforts would be necessary in the event of a climate change. However, the model indicates that in regions of high endemicity the use of insecticides and drugs may lead to an increase in incidence due to enhanced resistance development. Projected climate change, on the other hand, may lead to a limited reduction of the occurrence of malaria due to the presence of a higher percentage of immune persons in the older age class. Given this observation, in order to retard the evolution of resistance, a combination of methods or drugs should be used, combined with a selective high dosage rate for those people or areas most vulnerable. Elements of a sustainable antimalarial policy in regions of high endemicity will furthermore need to rely upon a stimulation of socio-economic development and provision of vector-proof housing. However, given the multiplicity of ecological and biological elements and of the natural, adaptive defence mechanisms of the malaria parasite/vector complex, control or eradication must be planned with the consideration of prevailing local conditions.

The modelling approach presented here fits in well with the qualitative attention currently being paid to the importance of evolutionary principles (e.g. Levy, 1992; Ewald, 1994). However, a great deal of empirical research is needed to improve the modelling approach. In the specific case of malaria, it is especially important that more insights into the possible shapes of the fitness functions of the parasites and the mosquitoes are acquired. This need is illustrated by the results on the impact of migration on the development of resistance at high doses, since they differ from the results of previous studies as a result of different assumptions regarding the fitness functions. Case studies on specific geographical regions need to be constructed to test the model against empirical data. For example, the fitness function of the mosquitoes and parasites as related to the use of insecticides and antimalarial drugs is not yet based on survey data, but rather on educated guesses.

Nevertheless, the fact remains that development of integrated assessment models which are based on the evolutionary and local dynamics of ecological systems may prove essential in assessing future developments in these complex adaptive systems. As adaptation to human intervention is a general phenomenon posing problems in health science (cf. Levy, 1992; Ewald, 1994), this approach might find wider applications in addressing resistance problems occurring in the management of other, re-emergent, infectious diseases (WHO, 1996).

Chapter 5

CLIMATE CHANGE, THERMAL STRESS AND MORTALITY CHANGES

INTRODUCTION

In healthy individuals, an efficient regulatory heat system enables the body to cope effectively with thermal stress. Within certain limits, thermal comfort can be maintained by appropriate thermoregulatory responses and physical and mental activities can be pursued without any detriment to health. Temperatures exceeding the comfortable limits, both in the cold and warm range, substantially increase the risk of (predominantly cardiopulmonary) deaths. Several mechanisms may explain this increased mortality: increased blood pressure, blood viscosity, and heart rate, associated with physiological adjustment to cold and warmth, may explain the temperature-induced increased mortality of diseases of the cardiovascular system (Keatinge et al., 1984; Pan et al., 1995). Indirectly, influenza contributes to cold-related mortality (Anderson & Le Richie, 1970; Kunst et al., 1993). Susceptibility to pulmonary infections may increase through bronchoconstriction, caused by the breathing of cold air (Schaanning, et al., 1986). Bull (1980) argues that excess mortality in winter may be due to physiological changes in cellular and humoral immunity, with behavioural factors also playing a role.

To determine the impacts of thermal stress on mortality associated with climate change, one needs an understanding of current and past relationships. However, a great deal of uncertainty surrounds the possible changes in mortality. A question that remains to be elucidated is the possible offsetting of the increase in heat-related mortality due to global warming by a decrease in cold-related mortality, also due to global warming. This balance between winter and summer mortality would to a large extent depend on local circumstances and adaptive responses, both physiological and behavioural, as well as technical.

Most of the recent research regarding the direct effect of climate change on heat-related morbidity and mortality has been related to the impacts of heatwaves (e.g. Kalkstein, 1993; see also McMichael et al., 1996). However, this chapter will not focus on periods of extreme heat or cold, but will consider the potential changes in numbers of deaths associated with warmth and moderate cold, related to the gradual influences of anthropogenic climate changes on health risk (Martens, 1997b). First, the scientific literature on the health impacts of moderate changes in long-term exposure to ambient levels of temperature will be reviewed and an aggregate dose-response relationship will be derived by means of a meta-analysis. Second, the dose-response relationship of the meta-analysis will be used to assess changes in mortality rates attributable to climate changes using projections for 20 cities throughout the world. The sensitivity of potential climate change-induced mortality changes to physiological adaptation and socio-economic developments will be tested.

HEALTH IMPACT ASSESSMENT

Central components of the health impact assessment of changes in temperature are the estimation of the temperature levels to which individuals are exposed, and the determination of the nature of the dose-response relationship for use in estimating levels of mortality changes due to a change in this environmental stressor.

Exposure Level: Climate Change Scenarios in Selected Cities

Climate conditions and physiological temperature thresholds of populations differ between locations. Although any selection of 20 cities is merely a reflection of the great diversity of urban environments and population behaviour (including rural areas), the cities used in this study represent a wide range of geographical locations, climate conditions, and levels of development (see Table 5.1), and broadly reflect these differences. The climate changes referred to in this study were constructed by the IPCC Working Group II: Impacts Assessment. Changes in documented current climate conditions were modelled according to the results of three transient GCMs: the GCM developed by the Max Planck Institute in Germany (ECHAM1-A) (Cubasch *et al.*, 1992), the GCM developed by the UK Meteorological Office (UKTR) (Murphy, 1995; Murphy & Mitchell, 1995) and the GCM developed at the Geophysical Fluid Dynamics Laboratory in the USA (GFDL89) (Manabe *et al.*, 1991, 1992). The baseline climatology relies on temperature data for the period 1961–1990, and is extracted from the NOAA baseline climatological data set (Baker *et al.*, 1994). The realised warming according to the three GCMs is 1.16°C, which lies close to the IPCC 'best estimate' around the year 2050 (Viner, 1994).

Table 5.1: Selected Cities

City (country; latitude-longitude)	Lowest/highest mean monthly temperature (°C)	City (country; latitude-longitude)	Lowest/highest mean monthly temperature (°C)
1. Mauritius (Mauritius; 20.43°S,57.67°E)	21.0 / 26.1	11. Athens (Greece; 38.00°N,23.44°E)	10.2 / 27.7
2. Buenos Aires (Argentina;.34.58°S,58.48°W)	11.1 / 24.5	12. Budapest (Hungary; 47.52°N,19.03°E)	0.3 / 20.9
3. Caracas (Venezuela;10.50°N,66.88°W)	20.4 / 23.2	13. London (UK; 51.52°N,0.12°W)	5.6 / 18.9
4. San Jose (Costa Rica; 9.93°N,84.08°W)	19.2 / 20.6	14. Madrid (Spain; 40.40°N,3.70°W)	6.0 / 24.2
5. Santiago (Chile; 30.30°S,70.40°W)	10.1 / 16.0	15. Zagreb (Croatia; 45.49°N,15.58°E)	1.4 / 21.0
6. Beijing (China; 39.93°N,116.28°E)	-4.3 / 25.9	16. Los Angeles (USA; 34.00°N,118.15°W)	14.2 / 23.9
7. Guangzhou (China; 22.13°N,113.32°E)	13.3 / 28.5	17. New York (USA; 40.45°N,74.00°W)	-0.2 / 24.3
8. Singapore (Singapore; 1.37°N,103.92°E)	25.7 / 27.4	18. Toronto (Canada; 43.42°N,70.25°W)	-5.0 / 21.8
9. Tokyo (Japan; 35.40°N,139.45°E)	5.2 / 27.1	19. Melbourne (Australia; 37.82°S,144.97°E)	10.2 / 20.9
10. Amsterdam (Netherlands; 52.30°N,4.77°E)	2.5 / 17.1	20. Sydney (Australia; 33.95°S,151.18°E)	12.2 / 22.6

The Effect of Temperature on Mortality: Literature Review

The impact of changes in climate on mortality rates has been addressed in numerous studies (for a survey of these studies, see, for example, Alderson (1985), and Green *et al.* (1994)). Several studies on heat mortality relationships assess the impact of periods of extreme temperatures (heatwaves) on mortality (e.g. Jones *et al.*, 1982; Katsouyanni *et al.*, 1993); others acknowledge the impact of variations of average temperatures on mortality (e.g. Rose, 1966; Dunningan *et al.*, 1970;

Table 5.2: Synopsis of Studies of the Effects of Temperature Exposition on Mortality

Health endpoint	Study area	Summary of findings	Study design	Reference
Ischaemic heart disease	Greater London 1970–1974; adults > 45 years	A linear association between number of IHD deaths at all ages with mean weekly temperature was observed. Proportional changes similar for all age groups.	Weekly time series	Bainton et al. (1977)
Total; stroke; arteriosclerotic heart disease	Ten metropolitan areas in the USA 1962–1965	Synoptic weather situations were associated with mortality changes: excessive heat and humidity and low wind speeds, and increased air pollution. Persons suffering from impairments of cardiovascular and respiratory system were most vulnerable.	Daily time series	Driscoll (1971)
Coronary heart disease	Scotland 1962–1966	Winter rise in incidence was related to environmental temperature; the spring rise was not related to temperature.	Monthly time series	Dunningan et al. (1970)
Total; heat-stroke	St. Louis and Kansas, USA; 1980	The July 1980 heatwave was associated with increased incidence of heat-stroke, especially among the elderly, the poor, non-whites and city dwellers. Overall mortality also increased during the heat wave.	Daily time series	Jones et al. (1982)
Deaths from all causes; deaths from weather sensitive causes.	Forty-eight cities in the USA; 1964–66, 1972–78 and 1980; all ages	The relationships between temperature and mortality were strongest in regions where hot weather is uncommon. Winter relationships were found to be weaker. Regional acclimatisation was found to be important, especially in the summer. A threshold temperature, beyond which deaths increase was identified for all causes and for both winter and summer. The elderly were more stressed than the other age groups.	Daily time series	Kalkstein & Davis (1989)
Total	Athens, Greece; 1981–87	The heatwave of July 1987 showed an increase in mortality rates as compared with the previous 6 years. The interaction between high temperatures and levels of sulphur dioxide was statistically significant and was suggestive for ozone and smoke.	Daily time series	Katsouyanni, et al. (1993)
Total	USA; 1921–1985	Warmer temperatures in the summer and colder temperatures in the winter were associated with higher mortality rates. Economic factors ameliorated the fatal impact of weather fluctuations.	Monthly time series	Larsen (1990)
Coronary heart diseases; cerebrovascular disease	Thirty-two metropolitan areas in the USA; 1962–1966	An inverse linear pattern of CHD and stroke mortality with temperature was observed, with low mortality rates at temperatures between 15.6°C–26.6°C, and then rising sharply at higher temperatures. Very hot days appeared to have a cumulative effect upon mortality. Snowfall was also found to be associated with higher mortality rates.	Daily time series	Rogot & Padgett (1976)
Coronary heart disease	Chicago, USA; 1967	An inverse linear association was found between daily temperatures and deaths due to CHD, although this relationship was not found for females. There were also more deaths on days with snowfall and with higher relative humidities.	Daily time series	Rogot (1974)
Coronary heart disease	England and Wales; 1950–1962	The winter excess of deaths was found to be highly correlated with coldness, but not with air pollution or rainfall. A fall in mortality with a rise in temperature was found over the whole of the observed temperature range (1.7°C–21.1 °C)	Monthly time series	Rose (1966)

Rogot, 1974; Rogot & Padgett, 1976; Bull & Morton, 1975, 1978; Bainton *et al.*, 1977; Shumway *et al.*, 1988; Larsen, 1990; Kunst *et al.*, 1993; Green *et al.*, 1994; Langford & Bentham, 1995; Pan *et al.*, 1995). Furthermore, several studies note that mortality can be further exacerbated by other meteorological factors such as wind, humidity, rainfall, and snow (e.g. Driscoll, 1971; Roberts & Lloyd, 1972; Rogot, 1974; Baker-Blocker, 1982; Kalkstein & Smoyer, 1993; Kunst *et al.*, 1993). A summary of some selected studies on a variety of climate-mortality relationships is given in Table 5.2.

Two important criteria were used to select the studies described below. Despite the great number of studies on the relationship between average temperature and mortality, only a limited number of these studies allow for a straightforward calculation of this relationship and its variance, which would allow for calculation of the confidence that can be ascribed to the effect estimated. Furthermore, for inclusion in the analysis, the temperature range between which the temperature-mortality relationship was examined should also be clearly defined. Mortality has always been a key health end point in epidemiological studies. It is a well defined health outcome and mortality data are routinely collected and are readily available for epidemiological analysis. The epidemiological studies are grouped by comparable health end points. The strongest associations with temperature are observed for cardiovascular and respiratory diseases. Some of the studies (on cardiovascular diseases) also provide a breakdown of mortality changes by age. The studies included in the analysis are shown in Table 5.3 and Figure 5.1A–C and are discussed below.[1]

Pan *et al.* (1995) showed that in Taiwan (1981–1991) a temperature-mortality relationship is especially important among the elderly. The range corresponding to the fewest deaths from coronary artery disease (26°C–29°C) and cerebral infarction (27°C–29°C) is higher than in countries with colder climates. Among elderly, the risk of cerebral infarction increases by 3.0 per cent per 1°C reduction from 27–29°C; below 26°C–29°C, the risk of coronary artery disease increases by 2.8 per cent per 1°C reduction; mortality from cerebral haemorrhage decreases with increasing temperature at a rate of 3.3 per cent per 1°C.

Highly significant associations between temperature and death from all causes, chronic bronchitis, pneumonia, ischaemic heart disease and cerebrovascular disease were found for England and Wales in the period 1968-1988 (Langford &

---

[1] After completion of this study, The Eurowinter Group (1997) published a paper which assessed increases in mortality per 1°C fall in temperature in various European regions. This study showed that mortality increased to a greater extent with a given fall in temperature in regions with warmer winters, in populations with cooler homes, and among people who wore fewer clothes and were less active outdoors.

Bentham, 1995). Based on their statistical analysis of mean monthly temperatures and monthly death rates, the change in mortality per 1°C increase between approximately -2°C and 20°C is estimated to be -2.1 per cent, -10.5 per cent, -3.4 per cent, -2.4 per cent and -2.8 per cent for deaths from all causes, chronic bronchitis, pneumonia, ischaemic heart disease and cerebrovascular disease, respectively.

A study in Israel (1976–1985; Green *et al.*, 1994) showed that monthly variations in ischaemic heart disease and stroke mortality are associated with changes in temperature and that the contributions of pneumonia and influenza are relatively small. The association is stronger in the older age groups. The study showed that even in semi-arid climates the magnitude of excess winter ischaemic heart disease and stroke mortality may be considerable.

Table 5.3: Studies Selected for the Estimation of the Effect of Temperature Changes on Mortality

Health end points	Study area and design	% Change in health endpoint per 1°C increase in average temperature change (95 % CI;s) temperature range	References
Cardiovascular diseases	**Combined**	**age <65:** **-1.6 (-2.8–0.4; 0.6) cold(a)** **0.7 (-0.1-1.5; 0.4) warm** **age >65:** **-4.1 (-5.5–2.7; 0.7) cold** **1.6 (-0.2-3.4; 0.9) warm**	
Coronary heart disease Cerebrovascular disease	England, Wales; 1968–1988; monthly time series	(b) -2.4 (-4.0--0.8;0.8) -2-20°C --2.8 (-4.6--1.0;0.9) -2-20°C	Langford & Bentham (1995)

Coronary heart disease
 	
Cerebral haemorrhage
	Taiwan; 1981–1991;
	daily time series
		(c)
		      9-26°C
		age 45-64: -3.0 (-6.5-0.0;1.6)
		age > 64: -2.8 (-5.1--0.1;1.3)
		      26-32°C
		age 45-64: 5.8 (0.0-13.8;3.5)
		age > 64: 3.7 (0.0-8.5;2.2)
		
		      9-28°C
		age 45-64: -4.3 (-7.0--2.1;1.3)
		age > 64: -3.5 (-5.9--1.5;1.1)
		      28-32°C
		age 45-64: -8.5 (-12.8--3.0;2.5)
		age > 64: -6.8 (-11.0--1.3;2.5)
			Pan *et al.* (1995)

Table 5.3 Continued.

Health endpoints	Study area and design	% Change in health endpoint per 1°C increase in average temperature change (95 % CI;s) temperature range	References
Cerebral infarction		9-29°C age 45-64: -1.8 (-6.6-1.6;2.1) age > 64: -3.8 (-6.5--1.3;1.3) 29-32°C age > 64: 22.0 (11.0-36.0;6.4)	
Coronary heart disease Cerebrovascular disease	Israel; 1976–1985; monthly time series	(d,e) 5-21°C age 45-64:- 9.3 (-10.8--7.1;0.9) age > 65:-10.6 (-11.8--9.8;0.5) age 45-64:-10.0 (-12.3--7.6;1.1) age > 65:-9.7 (-11.5--8.0;0.9)	Green *et al.* (1994)
Cardiovascular diseases	The Netherlands; 1979–1987; daily time series	(f) -1.2 (-2.4-0.0;0.6) < 16.5°C 1.1 (-0.1-2.3;0.6) > 16.5°C	Kunst *et al.* (1993)
Coronary artery disease	England, Wales; 1969–1971; monthly time series	(f) -2.5 (-5.4-0.4) < 17.9°C	West and Lowe (1976)

Coronary heart disease (myocardial infarction)
 	
Cerebrovascular disease
	England, Wales; 1963–1966 New York; 1965–1968; daily time series	(d,f)
		England and Wales: < 20°C
		age < 60: -1.2 (-2.4-0.0;0.6)
		age > 60: -1.9 (-3.9-0.1;1.0)
		
		New York: < 20°C
		age < 60: -0.9 (-1.9-0.1;0.5)
		age > 60: -1.0 (-2.0-0.0;0.5)
		                    > 20°C
		age < 60: 0.5 (-0.1-1.1;0.3)
		age > 60: 8.7 (0.1-17.3; 4.4)
		
		England and Wales: < 20°C
		age < 60: -1.3 (-2.7-0.1;0.7)
		age > 60: -2.0 (-4.0-0.0;1.0)
		
		New York: < 20°C
		age < 60: -0.6 (-1.2-0.0;0.3)
		age > 60: -1.7 (-3.5-0.1;0.9)
		                    > 20°C
		age < 60: 2.8 (0.1-5.5;1.4)
		age > 60: 15.1 (0.0-30.1;7.7)
			Bull & Morton (1975, 1978)

Cerebrovascular disease	New York; 1959–1963 Tokyo; 1960–1964 London; 1960–1964; seasonal time series	(f,g) New York: -0.8 (-1.8-0.2;0.4) 4-22°C Tokyo: -2.4(-5.3-0.5;1.2) 4-22°C London: -3.1(-6.8-0.6;1.6) 4-16°C	Sakamoto-Momiyama & Katayama (1971)

Table 5.3 Continued.

Health endpoints	Study area and design	% Change in health endpoint per 1°C increase in average temperature	References
		change (95 % CI;s) temperature range	
Total mortality	**Combined**	**-1.0 (-2.0-0.0; 0.5) cold (a)** **1.4 (-0.8-3.6; 1.1) warm**	
	England, Wales; 1968–1988; monthly time series	(b) -2.1 (-3.5--0.7;0.7) -2-20°C	Langford & Bentham (1995)
	The Netherlands; 1979–1987; daily time series	(f) -0.9 (-2.0-0.2;0.5) < 16.5°C 1.9 (-0.4-4.2;1.0) > 16.5°C	Kunst *et al.* (1993)
	Los Angeles; 1970–1979; daily time series	(h) -0.5 (-1.4-0.3;0.5) 10-23.3°C 0.6 (-1.9-3.0;1.3) 23.3-36.7°C	Shumway *et al.* (1988)
Respiratory diseases	**Combined**	**-3.8 (-6.9--0.7; 1.6) cold(a)** **10.4 (0.0-20.8; 5.3) warm**	
Chronic bronchitis	England, Wales; 1968–1988; monthly time series	(b) -10.5 (-17.2--3.8;3.4) -2-20°C	Langford & Bentham (1995)
Pneumonia		-3.4 (-5.6--1.2;1.1) -2-20°C	
Respiratory diseases	The Netherlands; 1979–1987; daily time series	(f) -2.7 (-5.4-0.0;1.4) < 16.5°C 10.4 (0.0-20.8;5.3) > 16.5°C	Kunst *et al.* (1993)
Pneumonia	England, Wales; 1963–1966; daily time series	(f) < 20°C age < 60: -5.4 (-10.9-0.1; 2.8) age > 60: -4.7 (-9.4-0.0; 2.4)	Bull & Morton (1975, 1978)

(a): Cold: average temperature below the comfort temperature; warm: average temperature above the comfort temperature as defined in text.

(b): Calculated as percentage difference between mean mortality at yearly average temperature between 1968–1988 (=6°C) and 1 °C increase. No influenza included and in the regression equation 'year' was set at 88. Standard error was estimated as in (f) with a p-value of 0.001.

(c): Calculated between the difference of the odds ratio at comfort temperature and 10°C below this temperature divided by 10 and the difference between the odds ratio at comfort temperature and the maximum temperature used in the study (32°C) divided by the difference in temperature. Effect estimate of cerebral haemorrhage not included in the combined estimate.

(d): Study results averaged.

(e): Calculated as mean ratio of winter to summer mortality divided by the difference in winter and summer minimum temperature (=15°C).

(f): The standard error was estimated on the basis of p-values and assuming a normal distribution, with a p-value of 0.05. The standard error was calculated based on the upper limit of the cutoff probability: $z = 1.65$ for a p-value of 0.05; standard normal distribution: $z = (x-\mu)/s$; H_0: $\mu = 0$; 95 per cent; confidence interval = $\mu \pm 1.96\ s$.

(g): Average of spring and autumn values.

(h): Calculated as percentage difference between mortality at most comfortable temperatures and mortality at lowest and highest temperatures, divided by the temperature difference between comfort, lowest and highest temperatures (using the weighted least-squares model and fixed CO_2 concentrations).

A study by Kunst *et al.* (1993) estimated the relationship between temperature and mortality rates due to a variety of causes of death in The Netherlands (1979–1987) by regression analysis. The relationship between warmth (temperature equal to or higher than 16.5°C) and mortality from respiratory disease is greater than that from cardiovascular diseases, and patients with less severe respiratory disease are also at risk. These deaths are not compensated for by lower death rates in subsequent weeks. Not only heat, but also moderately high temperatures are related to mortality rates. The study also found that humid weather is protective against the effects of warmth, which suggests that the inhalation of dry and warm air is particularly harmful to the respiratory system. The effects of cold (temperatures below 16.5°C) are greater for cardiovascular diseases than for total mortality. The following relationship was arrived at, controlled for influenza, sulphur dioxide (SO_2) density and season: the percentage change in total mortality associated with a 1°C increase in the average value of cold or warmth for total mortality was found to be 0.9 per cent and 1.9 per cent. Below a mean temperature of 16.5°C, mortality rates decrease by 1.2 and 2.7 per cent per degree Celsius increase for cardiovascular and respiratory diseases, respectively. Above 16.5°C, mortality rates increase by 1.1 and 10.4 per cent per degree Celsius increase. Respiratory diseases are thus more sensitive to changes in temperatures.

Shumway *et al.* (1988) applied a non-linear model of temperature and mortality to a data set of Los Angeles (1970–1979). Above 23.3°C, total mortality increases by 0.6 per cent per degree Celsius. Non-significant contributors to mortality are sulphur dioxide, nitrogen dioxide, ozone and relative humidity.

Bull & Morton (1975, 1978) found that deaths from myocardial infarction, strokes and pneumonia fall linearly as temperatures rise between -10°C and 20°C (based on data from England and Wales for the years 1963–1966 and New York for the years 1965–1968). Below -10°C and above 20°C death rates rise sharply as temperature decreases or increases, respectively. No differences were found between males and females and a more pronounced effect was noticed in the elderly. Mortality rates due to pneumonia decrease by about 5 per cent per degree Celsius increase between -10°C and 20°C. Death rates due to myocardial infarction and stroke decrease by about 1 per cent. Each degree Celsius increase above 20°C causes cardiovascular mortality rates to increase by between 0.5 and 15.1 per cent, depending on age and disease.

Also based on data from England and Wales (1969–1971), West & Low's (1976) study found ischaemic heart disease to be highly correlated with temperature, rainfall and a socio-economic index. Linear regression analysis indicates that the monthly mortality from ischaemic heart disease increases by approximately 2.5per cent for a 1°C drop in mean monthly temperature.

Sakamoto-Momiyama & Katayama (1971) performed a quantitative analysis of the geographical difference in the relationship between temperature and mortality in the three cities of London, Tokyo and New York (1959–1964). Mortality appears to be inversely related to temperature, and stroke increases at low temperatures (up to about 22°C) by 0.8 per cent, 2.4 per cent, and 3.1 per cent with a 1°C decrease in temperature for New York, Tokyo, and London respectively. The smallest changes in mortality in New York indicate that central heating has had a favourable effect on mortality.

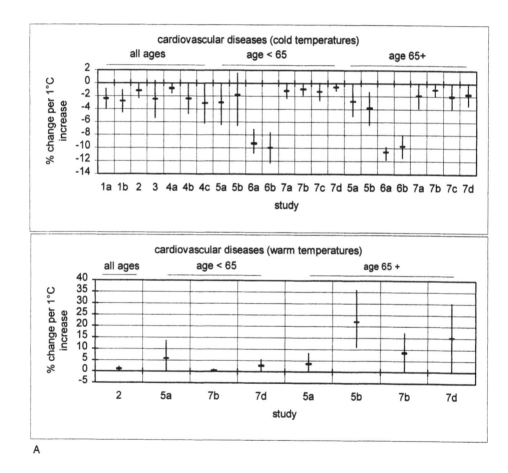

A

Figure 5.1 A : Change in Mortality due to 1°C Increase in Average Temperature (95 Per Cent Confidence Interval) (See Caption Figure 5.1 B+C)

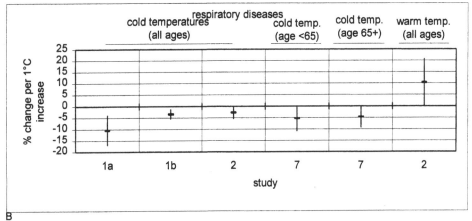

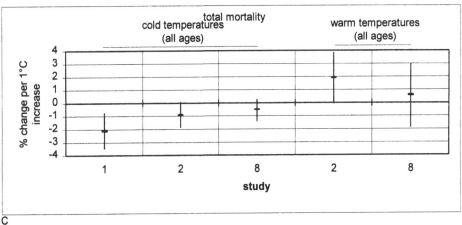

Figure 5.1 B+C: Change in Mortality due to 1°C Increase in Average Temperature (95 Per Cent Confidence Interval): Study: 1: Langford & Bentham (1995); 2: Kunst *et al.* (1993); 3: West & Lowe (1976); 4: Sakamoto-Momiyama & Katayama (1971); 5: Pan *et al.* (1995); 6: Green *et al.* (1994); 7: Bull & Morton (1975,1978); and 8: Shumway *et al.* (1988);

A. Cardiovascular mortality:

1a: Coronary Heart Disease, b: Cerebrovascular Disease; 2: Cardiovascular Disease; 3: Coronary Heart Disease; 4a-c: Cerebrovascular Disease (a: New York, b: Tokyo, c: London); 5a: Coronary Heart Disease, b: Cerebral Infarction; 6a: Coronary Heart Disease, b: Cerebrovascular Disease; 7a-b: Myocardial Infarction, c-d: Cerebrovascular Disease (a+c: England and Wales, b+d: New York);

B. Respiratory Mortality:

1a: Chronic Bronchitis, b: Pneumonia; 2: Respiratory Diseases; 7: Pneumonia.

C. Total Mortality

Effect Estimate

Meta-analysis is a method for quantitatively combining the measures of effects across studies. To provide a combined estimate of the effect of a 1°C change in temperature for each health end point, a weighted average was calculated. To give more emphasis to studies that had lower errors in their effect estimates, the study-specific results presented in Table 5.3 were weighted by their inverse variance (reciprocal of the square of the standard error). The standard error of this central estimate was obtained by weighting the standard errors by the inverse of their square. It should be noted that the majority of the studies referred to developed cities in cold-temperate and temperate climates, describing the relation between mortality and low temperatures; very few studies which described the relation between warmth and mortality were available in the literature. Obviously, the selection of studies included in the analysis ultimately determines the final outcome in the meta-analysis, which, in this case, is slightly biased towards studies of cold-related mortality.

The combined effect estimates are given in Table 5.3. The relationship between mortality and temperature is visualised as a V-shaped function, with mortality not only rising at very high and low temperatures, but also at more moderate temperatures (see Figure 5.2). Further, this relationship is remarkably consistent from study to study (i.e. from area to area) (although some studies suggest that the relationship between temperature and mortality is J-shaped, to indicate a steeper slope at higher temperatures; see McMichael *et al.* (1996)). The main difference between areas is the exact temperature at which mortality is lowest (i.e. the comfort temperature). This geographical difference in comfort temperature reflects not only socio-economic conditions, but also physiological adaptation. The comfort varies between 16.5°C for The Netherlands (Kunst *et al.*, 1993) and 29°C for cerebral infarction in Taiwan (Pan *et al.*, 1995). According to Bull & Morton (1978) deaths from myocardial infarction and stroke in New York increase at temperatures above about 20°C, and a rather similar finding was reported by Rogot and Padgett (1976) in the USA and by Näyhä (1980) in Finland. Shumway *et al.* (1988) found a comfort temperature of about 23°C for Los Angeles (1970–1979); Rogot & Blackwelder (1970) reported approximately 24°C as the comfort temperature for Memphis (USA). Since people adapt (physiologically) to climate, a comfort temperature of 25°C for Mauritius and Singapore was used (based on current climate conditions). For Amsterdam and Los Angeles, the comfort temperatures chosen were 16.5°C and 23°C, respectively. For the other cities, the comfort temperature was set at 20°C. Although the assumption that the comfort

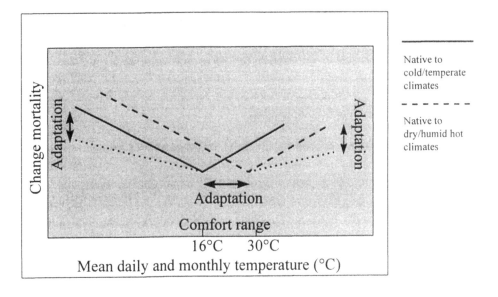

Figure 5.2: V-Shaped Relationship between Mean Daily/Monthly Temperature and Mortality; βcold and βwarm Represent the Slopes of Curve

temperature is somewhat higher for warmer areas seems reasonable, the sensitivity of the results to this assumption is tested in the next section.

Three studies provided a breakdown of cardiovascular mortality by age: based upon these, the effect of a 1°C increase was estimated for the age group younger than 65 and the age group 65 or older. For total and respiratory mortality, the studies included provided no clear differences in the temperature effects on mortality between these two age groups. Furthermore, a distinction was made between the effects of an increase in temperature when the average temperature is 'cold' (i.e. temperature below the comfort temperature: βcold) or 'warm' (i.e. temperature above the comfort temperature: βwarm). Although the comfort temperatures varied in the studies considered, the changes in mortality due to an increase in 'cold' and 'warm' temperatures were combined.

For total mortality, the weighted effect estimate for an increase of 1°C in the 'cold' range is -1.0 per cent. This estimate is -3.8 per cent for respiratory diseases and -1.6 per cent and -4.1 per cent for cardiovascular diseases (for the age group below and above 65, respectively). In the 'warm' temperature range, the weighted mean estimate of an increase of 1°C for total mortality is 1.4 per cent. For cardiovascular diseases, these estimates are 0.7 per cent and 1.6 per cent for the age group below and above 65, respectively; for respiratory disease the weighted effect estimate is 10.4 per cent increase per 1°C increase. Using the data from the studies described above, total mortality and respiratory disease seem to be more sensitive to a 1°C increase in the 'warm' temperature range than a 1°C increase in the 'cold' temperature range, in contrast to cardiovascular diseases. However, for total and respiratory mortality the estimates are based on only a few studies, and must therefore be interpreted with caution.

THERMAL STRESS IN A NUMBER OF CITIES

The potential effects of climatic changes on mortality changes due to thermal stress have been estimated using several assumptions. First, the coefficients of warmth and cold-related mortality have been applied straightforwardly to the change in the cities' temperatures. Second, the city-specific estimates have been aggregated to yield estimates of changes in countries' cardiovascular mortality rates. Finally, the sensitivity of these outcomes to physiological as well as socio-economic adaptation (i.e. changes in comfort temperature, and strength of the cold- and warmth-related changes) is examined.

Changes in mortality rates due to thermal stress relative to 1990 have been estimated by combining the changes in the cities' monthly mean temperatures projected by the three GCMs with the dose-response relationship derived from the meta-analysis. Figure 5.3A–C shows changes in yearly averaged total, cardiovascular and respiratory mortality, for the three GCMs.

In applying the temperature mortality coefficients straightforwardly – i.e. taking into account both 'cold' and 'warmth' effects on mortality – the cities included in this study can be broadly classified into three categories in respect of the mortality changes due to an anthropogenic climate change. In cities with present temperatures close to the comfort temperature for all months (e.g. Caracas, San Jose, Singapore) mortality rates (total, cardiovascular and respiratory) increase for most of the climate change scenarios employed. However, decreases in winter mortality may offset excess summer mortality in a second group of cities with relatively colder climates, such as Santiago, Amsterdam, Budapest, London, Zagreb, Toronto and Melbourne. Finally, reflecting the magnitude of the mortality coefficients (the higher sensitivity of total and respiratory mortality to a 1°C increase above the comfort temperature compared to below this temperature), the other cities in this study, where temperatures exceed the comfort temperature for several months, may experience an increase in total and respiratory mortality rates, in contrast to a decrease in cardiovascular disease mortality. People above the age of 65 are more 'temperature-sensitive' than younger people.

The differences shown in Figure 5.3A–C not only indicate the importance of cities' current climate conditions in the effect estimate, but also illustrate the large interregional differences and those between GCMs with respect to the climate

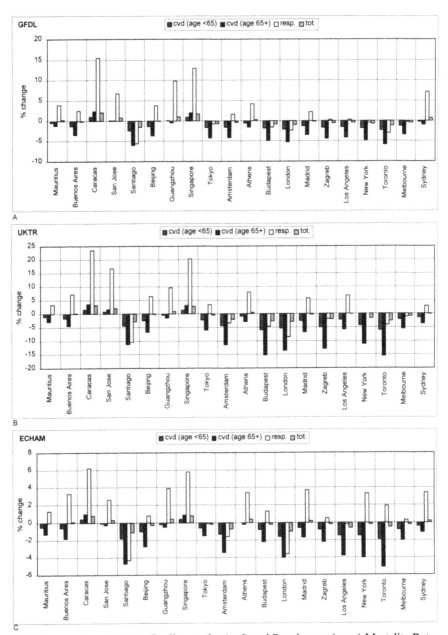

Figure 5.3: Changes in Total (tot.), Cardiovascular (cvd) and Respiratory (resp.) Mortality Rates for the 20 Selected Cities due to Thermal Stress. A: GFDL89-GCM; B: UKTR-GCM; C: ECHAM1A-GCM

change scenarios. Estimates of the average mortality change may vary between about -2 per cent (Santiago) and ~2 per cent (Caracas), and between ~-7 per cent (Santiago) and ~15 per cent (Caracas), for total and respiratory mortality, respectively. For cardiovascular disease mortality this varies between ~-3 per cent (Toronto) and ~1 per cent (Caracas), and between ~-9 per cent (Toronto) and ~2 per cent (Caracas), for the age group below and above 65, respectively. It should be noted again that the effect estimates between temperature and total and respiratory mortality are based on only a few studies, and therefore have to be interpreted with caution. Furthermore, the outcomes presented are sensitive to changes in comfort temperature, and the magnitude of the temperature-mortality coefficient. For cardiovascular mortality, where the estimate of the temperature-mortality relationship is based on more empirical data and is therefore more robust, this sensitivity is examined in the next section.

CARDIOVASCULAR MORTALITY AND SENSITIVITY TO ADAPTATION

People may adapt physiologically to warm temperatures. Acclimatisation may occur in several days, although complete acclimatisation may take up to several years (Babayev, 1986). Figure 5.4 shows the effect of different assumptions regarding the V-shaped relationship between temperature and cardiovascular mortality, with the temperature baseline for Amsterdam used as an example; the temperature in each month is increased by 0°C–3.5°C. In Figure 5.4A the comfort temperature for the cities' populations (16.5°C) is varied in the range -2°C to +2°C.

Clearly, the change in mortality rates is the result of the value of the comfort temperature, the temperature increase and the mortality coefficients. However, even lowering the 'ideal temperature' for the selected city by 2°C will still, on aggregate, lead to a decrease in mortality rates due to cardiovascular diseases as temperatures increase by 3.5°C. Although it is hardly possible to estimate how physiological adaptation may take place in the various populations, the figures show that any adaptation to warmer climates (i.e. a positive deviation from the comfort temperature) will result in a higher decrease in the mortality rates in cities with lower temperatures (i.e. cities of the second group as described before) and a lower increase in mortality in the cities belonging to the first group.

Besides physiological adaptation, cultural adjustments through time may have an impact on the relationship between climate and mortality rates. In developed countries, socio-economic progress over time has led to a reduction in winter excess mortality (e.g. Momiyama and Katayama, 1972; Näyhä, 1980; Keatinge *et*

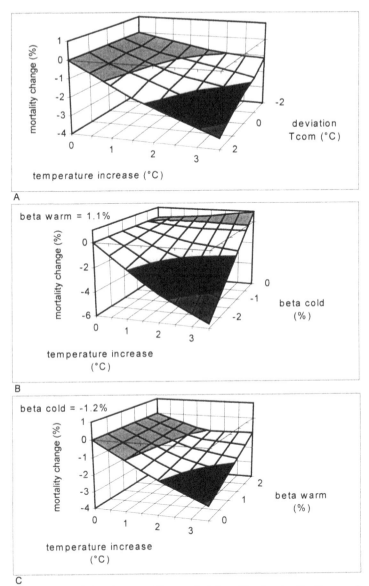

Figure 5.4: Sensitivity of the Average Change in Mortality due to Cardiovascular Diseases, Using the Baseline Temperature of Amsterdam and the V-Shaped Relationship Described in Kunst *et al.* (1993), to Changes in: (A) The Comfort Temperature (i.e. Temperature at which Mortality is Lowest); (B): βcold (βwarm is Fixed at 1.1 Per Cent); and (C): βwarm (βcold is fixed at -1.2 Per Cent); Coefficients Vary Between Zero and Central Estimate Plus Twice the Standard Deviation (See Table 5.3); Monthly Mean Temperatures are Increased Between 0 and 3.5°C

al., 1989), not only due to improvements in housing conditions, but also as a result of better clothing and the wider availability of food and fuel (Kunst *et al.*, 1993). Air conditioning may also be a mitigating factor, though this is much less likely to be a confounder in developing countries (Kalkstein, 1993). Furthermore, there is evidence that people living in poverty, as well as urban populations in developing countries, are particularly vulnerable to thermal change. Poor housing conditions, including the lack of access to air conditioning, as well as the so-called 'urban heat island' effect, are among the main causes (Kilbourne, 1989). In Figure 5.4B–C the sensitivity of cardiovascular mortality changes due to socio-economic progress is illustrated by changing βcold (with fixed βwarm). The increasing possibilities of mitigating the effects of warmer temperatures (e.g. air conditioning) may be reflected by a decrease in βwarm (with fixed βcold).

Table 5.4 presents estimates of changes in numbers of people dying due to cardiovascular mortality, with the temperature change being the average of the three GCMs considered. Changes in the baseline (1985–1989) country-specific mortality figures (WHO, 1993b, 1995b) are based on the city-specific estimates. Although for relatively small countries, such as Mauritius, Costa Rica and The Netherlands, the temperatures in the corresponding cities may be a good approximation of the country's climate, this will often not be the case for larger countries. Furthermore, in rural areas the relative importance of temperature for mortality may be different from the effect on urban populations. The results presented in Table 5.4 should therefore be seen as an indication of possible changes and not as a prediction as such.

Table 5.4 shows that increasing temperatures clearly have the greatest effect on the oldest age group (either positive or negative). This is not only due to the higher 'temperature sensitivity' of the old, but also to the (much) higher baseline mortality of this age group. Although the balance will vary by country, the number of cardiovascular deaths avoided associated with a global warming of ~1.2°C, keeping other things constant, may be as much as about 300/100,000 in the oldest age group. However, a continuing decreasing trend in winter mortality would tip the scales in these colder regions to increasing excess summer mortality rates (Figure 5.4B). The change in mortality in cities with high temperatures during all months is rather insensitive to changes in βcold. With no decreasing winter mortality rates, global climate change may induce an excess of up to ~50 deaths per year per 100,000 people above 65. Adaptation to warmer climates, whether physiological (Figure 5.4A) or technical/behavioural (Figure 5.4C), will reduce this figure, the effect of adaptation being more pronounced in 'warm cities'.

Table 5.4: Baseline Cardiovascular Mortality and Estimates of Changes due to Thermal Stress

Country	Baseline mortality[a] (/100,000 pop)		Temperature-related mortality changes (/100,000 pop)		Cold-related mortality changes (/100,000 pop)		Warmth-related mortality changes (/100,000 pop)		Total temperature related mortality changes (/100,000 total population)
	age<65	age65+	age<65	age65+	age<65	age65+	age<65	age65+	
Mauritius	264	3737	-2	-70	-3	-100	1	30	-3
Argentina	192	3012	-2	-100	-4	-139	1	38	-10
Venezuela	97	2067	1	48	0	0	1	48	3
Costa Rica	64	2147	0	8	0	-24	0	31	0
Chile	80	2167	-2	-159	-2	-159	0	0	-11
China[b]	-	2510	-	-67	-	-103	-	37	-4
Singapore	121	2132	1	43	0	0	1	43	3
Japan	76	1828	-1	-61	-1	-79	0	18	-8
Netherlands	102	2441	-3	-163	-3	-181	0	19	-23
Greece	118	3011	-1	-51	-1	-86	1	35	-7
Hungary	291	4432	-9	-342	-10	-380	1	38	-49
UK	167	3036	-5	-240	-5	-250	0	10	-41
Spain	95	2405	-1	-96	-2	-129	1	33	-13
Croatia[c]	191	4159	-5	-285	-6	-324	1	39	-28
USA[b]	135	2675	-3	-152	-4	-184	1	32	-20
Canada	99	2323	-3	-209	-4	-235	1	26	-27
Australia[b]	110	2707	-1	-76	-2	-98	0	22	-9

(a): Based on WHO (1993b), except for China (for China no data for age group < 65 (-); WHO (1995b)).
(b): Effect estimate average of two cities included.
(c): Data from former Yugoslavia.

DISCUSSION AND CONCLUSIONS

The temperature-mortality effect estimates are based on a variety of studies, mostly conducted in developed, non-tropical countries, some of them several decades ago. Part of the spread of the point estimates of the individual studies reflects more than sampling variations, length-of-record problems and other difficulties inherent in studies developed before masses of digital mortality and weather data were available. Different populations, with different characteristics and circumstances, respond differently to specified temperature changes, which is also illustrated by the lower excess winter mortalities in Scandinavian countries compared with the UK (Sakamoto-Momiyama, 1978). Therefore, the use of one aggregate temperature-mortality relationship does not imply that this is a 'universal' dose-response relationship, but is intended to shed light on the effect of climate change on the balance between cold- and warmth-related mortality changes. However, although the estimates of the individual studies are sensitive to regression specification and the sample of cities/regions under investigation, the results are consistent.

The results suggest that climate change in areas with temperate, particularly cold temperate climates could result in a reduction of the disease burden due to less excess winter cardiovascular mortality, especially in elderly people. Based on current mortality levels, this decrease could be about 50 people per 100,000 population. A similar result is found in Langford & Bentham (1995) and Alderson (1985). The increase in cardiovascular mortality in warmer climates may be about 3/100,000. For total and respiratory mortality the results are more difficult to interpret. A great part of the overall winter mortality is due to infectious diseases, such as influenza, which depend on aerosol transmission (usually in places with poor ventilation). This infectious disease risk may be affected by an increase in the winter temperature of a few degrees if it were to encourage people to spend more time outdoors. However, it is unlikely that a small rise in temperature would drive people out of doors in the winter, meaning that infectious disease contact and respiratory distress would be similar at both temperatures. However, after controlling for influenza, some of the studies still show a strong relationship between cold temperatures and total and respiratory mortality (e.g. Kunst *et al.*, 1993; Langford & Bentham, 1995), suggesting that an increase in winter temperature does have an effect on mortality rates.

Furthermore, an anthropogenic climate change is likely to increase the frequency or severity of heat waves. It should be emphasised that, although this study focuses on the long-term influence of climate changes upon health risk and can not be directly compared with studies focusing on the effects of climate change on marked short-term fluctuations in weather, research on heatwave-related mortality suggest an increase in predominantly cardiorespiratory mortality and illness (McMichael, 1996).

The changes in mortality presented in this chapter will occur against a background of changes in levels of incidence of cardiovascular and respiratory diseases due to changes in prevailing lifestyles (e.g. diet, smoking behaviour), age distributions, levels of air pollution, etc. Developing countries are likely to be more vulnerable due to rapid urbanisation, few social resources and a low pre-existing health status.

In summary, this study has tried to answer the question: "What is the annual balance between changes in moderate cold- and warmth-related deaths due to global climate change, in different geographical and population settings?" Although the overall balance remains difficult to quantify and would depend on adaptive responses and existing health levels, global warming may cause a decrease in mortality rates, especially of cardiovascular diseases. Since it is based on empirical data from a variety of settings, this work should be extended by inclusion of data from a wider range of populations to further test the model's robustness.

Chapter 6

THE IMPACT OF OZONE DEPLETION ON SKIN CANCER INCIDENCE

INTRODUCTION

Depletion of the ozone layer has become a topic of considerable importance and concern. Evidence indicates that the concentration of stratospheric ozone has already decreased globally over the past several years (Stolarski *et al.*, 1991) and additional decreases are to be expected over the coming decades. A further depletion of the ozone layer would lead to an increase in the amount of UV radiation; UV radiation can be divided into UV-A (wavelengths 315–400 nm), UV-B (280–315 nm), and UV-C (100–280 nm). If short-wave UV radiation is absorbed by various components of plant and animal cells, several essential molecules are damaged, most notably the DNA which carries the genetic instructions for the synthesis of proteins, fundamental to a proper functioning of the cell. As all living organisms are built up of cells, solar UV radiation may be expected to have a widespread and fundamental effect on our biosphere, including direct effects on man (de Gruijl & van der Leun, 1993). The adverse effects, particularly of UV-B radiation, on human health are: sunburn, skin cancer, skin ageing, cataracts, and a probable impairment of human resistance to infectious diseases. The effects of excessive levels of UV-B on agriculture and ecosystems are: reduced crop yields, damage to phytoplankton, and adverse effects on aquatic food chains (UNEP, 1994).

This chapter focuses on the changes in the incidence of skin cancer associated with stratospheric ozone depletion in The Netherlands and Australia (Martens *et al.*, 1996). Two main categories of human skin cancer can be distinguished: (a) cutaneous malignant melanoma skin cancer (MSC), originating from pigment cells (melanocytes) in the skin; and (b) non-melanoma skin cancer (NMSC), originating from the normal epithelial cells (keratinocytes) in the outermost layer of viable skin (the epidermis). The latter can be subdivided into two main types: the basal cell carcinoma (BCC, most common) and the squamous cell carcinoma (SCC).

Skin cancers are among the most frequently occurring cancers in The

Netherlands; in Australia skin cancer incidences are the highest reported in the world (Giles *et al.*, 1988). People generally get sunburned during the first few days of bright sunlight, and overexposure occurs mostly during weekend outings or while working in gardens. Accumulated UV damage to the skin is also very common in outdoor workers. Australia has a very sunny climate, the majority of the inhabitants are of Caucasian origin, and the country is situated close to the Antarctic with its 'ozone hole', so the problem of ozone depletion and the related increases in ground-level UV radiation is of great concern. Although The Netherlands is located in a temperate climate zone, the UV radiation from the (summer) sun has notable health effects: people generally get sunburned in early spring during the first few days of sunlight, and accumulated UV damage to the skin is also very common in outdoor workers.

The analysis of what happens to the tumour incidences in the course of time after the ozone layer changes is complex, in particular due to the relatively long 'incubation' time between initial UV exposure and the first appearance of cancer. In this chapter an adapted and extended version of the UV-B chain model, first described in Slaper *et al.* (1992) and den Elzen (1993), is used to assist in assessing the changes in skin cancer incidences in relation to changes in the amount of stratospheric ozone, for The Netherlands and Australia. (Recently, a similar study for the USA and north-west Europe has been published (Slaper *et al.*, 1996).) This assessment model integrates dynamic aspects of the full source-risk chain: production and emission of ozone depleting substances, global stratospheric chlorine concentrations, local depletions of stratospheric ozone, the resulting increases in UV-B levels, and finally, the effects on skin cancer rates. This contrasts with earlier skin cancer assessments, which were based on the comparison of two stationary situations (Madronich & de Gruijl, 1993), and did not include the delay between exposure and tumour development (Kricker *et al.*, 1994). As future skin cancer rates depend on changes in the rate of stratospheric ozone depletion, changes in human exposure habits, the sensitivity of the population at risk, and the age distribution and size of the population, model simulations are performed with various assumptions with regard to these factors.

MODELLING THE CAUSE–EFFECT CHAIN

Stratospheric Ozone Depletion

Because stratospheric ozone depletion involves a combination of a multitude of causes, any simple analytical description is necessarily an incomplete one. For the purposes of this study, annually averaged latitudinal stratospheric ozone losses are

---

Box 6.1:
Stratospheric Ozone Depletion

Stratospheric ozone depletion is expressed as:

$$\Delta O(\phi, t) = k(\phi)[Cl(t) - Cl_0] \qquad (6.1)$$

where $\Delta O(\phi, t)$ is the annually averaged relative change in total column ozone at the $10°$ latitude band at time t (per cent/year). The mid-latitude of the corresponding band ϕ is equal to $55°N$ for The Netherlands; because of the size of the continent, Australia is divided into three latitude zones (ϕ zone 1 is $15°S$, ϕ zone 2 is $25°S$ and ϕ zone 3 is $35°S$; t varies over the period 1978–2050, and $Cl(t)$ is the global stratospheric chlorine concentration at time t (parts per million by volume (ppbv)), while Cl_0 is the global stratospheric chlorine concentration in the late 1970s, during which no depletion of the ozone layer is assumed to have occurred (= 1.9 ppbv in 1978).

Finally, $k(\phi)$ is the latitudinally dependent coefficient (per cent/ppbv) based on the calibration of the trend in the simulated stratospheric chlorine concentration against the ozone trend over the period 11/1978–05/1991 based on TOMS (total ozone mapping spectrometer) data. The ozone trend is -0.35 $\pm$0.14 per cent/year at $\phi(55°N)$, and about -0.02$\pm$0.14 per cent/year, -0.11 $\pm$0.14 per cent/year, and -0.29$\pm$0.14 per cent/year for the three zones of Australia (de Winter-Sorkina, 1995). This, with a trend in chlorine concentration of 0.11 ppbv/year, results in the latitudinal coefficients ($\pm$ 2 standard deviation): $k(55°N)$ at -3.2 $\pm$1.3 per cent/ppbv, $k(15°S)$ at -0.2 $\pm$1.3 per cent/ppbv, $k(25°S)$ at -1.0 $\pm$1.3 per cent/ppbv, and $k(35°S)$ at -2.6 $\pm$1.3 per cent/ppbv.

The initial stratospheric ozone abundance at month m (when no ozone depletion has occurred; $O_{0,m}$ (in Dobson units or DU)) is described by means of a sine function, taking into account the natural seasonal variation of the ozone column:

$$O_{0,m}(\phi) = O_{ref}(\phi) + O_A(\phi)\cos((2\pi / 12)(m - \eta(\phi))) \qquad (6.2)$$

For the purposes of the model, the reference value of the yearly averaged thickness of the ozone layer at the latitude under consideration ($O_{ref}(\phi)$) has been given the value of 330 (DU) for The Netherlands, and a value of 265, 285, and 317 DU for zones 1, 2 and 3 of Australia, respectively. The values of the amplitude describing the variation in the thickness of the ozone layer ($O_A(\phi)$) are 50, 9, 21, and 40 DU for The Netherlands and the Australian zones 1-3 (Bowman, 1985; Willems & Koken, 1995). $\eta(\phi)$ is the phase shift, i.e. the month number that has the maximum thickness of the ozone layer (April for The Netherlands, October for zone 1 of Australia and September for the other two zones).

The monthly thickness of the ozone layer ($O_m(\phi, t)$) at latitude ϕ at time t is now calculated as a linear function of the annual latitudinal change in the thickness of the layer of stratospheric ozone $\Delta O(\phi, t)$ (equation 6.1) and the initial monthly stratospheric ozone levels ($O_{0,m}(\phi)$), as:

$$O_m(\phi, t) = O_{0,m}(\phi)(1 + c_m(\phi)\Delta O(\phi, t)) \qquad (6.3)$$

Here $c_m(\phi)$ represents monthly factors at latitude ϕ correcting for monthly variation in the ozone depletion, in the light of TOMS data (de Winter-Sorkina, 1995; see Figure 6.2).

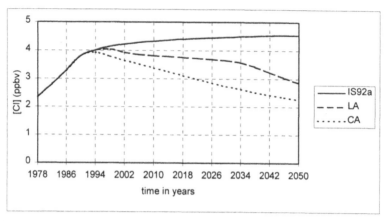

Figure 6.1: Tropospheric Chlorine Concentration for the IPCC IS92a Scenario, and for the London (LA) and Copenhagen (CA) Amendments to the Montreal Protocol

assumed to be linearly and directly dependent on stratospheric chlorine concentrations (see Box 6.1). The major source of uncertainty inherent in such an approach is the extremely simplified description of the atmospheric chemistry and transport processes in the lower stratosphere, although model studies do indeed show that the ozone budget in the lower stratosphere is mainly controlled by the evolution of the chlorine oxide chemistry, and is less affected by nitrogen oxides or bromine-related chemistry (cf. Granier & Brasseur, 1992).

Stratospheric chlorine levels are calculated using the halocarbon module (den Elzen, 1993) of the UV-B chain model (Slaper *et al.*, 1992). This model describes the cause-effect relationships underlying the stratospheric ozone depletion problem: consumption policies vis-à-vis CFCs and other halocarbons and global stratospheric chlorine concentrations. Historical data on CFC production have been used up to 1990. Three scenarios were used for the period 1990–2050: the IPCC IS92a scenario, the London Amendments, and the Copenhagen Amendments to the Montreal Protocol (see Figure 6.1). The control measures set out in the London Amendments to the Montreal Protocol (1987) provide for the phasing out of the consumption of CFCs, halons and carbontetrachloride (CCl_4), by the year 2000, and methylchloroform (CH_3CCl_3) by 2005.

In 1992 in Copenhagen an international agreement was adopted to phase out all of the regulated halocarbons as early as 1996. The phasing out of the alternatives

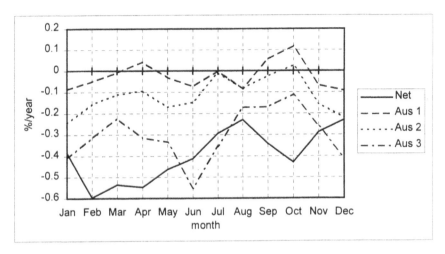

Figure 6.2: TOMS Total Ozone Monthly Trends of the Period 11/1978–06/1991 (de Winter-Sorkina, 1995), for The Netherlands and the Three Australian Latitude Zones

(hydrochlorofluorcarbons (HCFCs)) would proceed in several steps, with a 35 per cent reduction of the maximum HCFC level by 2005, a 65 per cent reduction by 2010, a 90 per cent reduction by 2015, and a total phase out by 2020. In the IS92a scenario, developed by the IPCC, 70 per cent of the developing world is assumed to ratify and comply with the London Amendments. Furthermore, in this scenario, it is assumed that most of the world develops and uses CFC substitutes, which will lead to a gradual phase-out of all CFC use from 2025 onwards, even in non-signatory countries.

Tropospheric chlorine concentration will manifest itself as lower stratospheric chlorine after a time delay of 2–4 years (in this study a delay of 4 years is assumed, cf. Prather *et al.* (1992)), and allowance is made for differences in absolute quantities between tropospheric and stratospheric concentration due to incomplete oxidation of the halocarbon source gases (see den Elzen, 1993). Recent analysis indicates that the tropospheric chlorine concentrations, attributable to anthropogenic halocarbons, peaked near the beginning of 1994, suggesting that the amount of reactive chlorine (and bromine) will reach a maximum in the stratosphere between 1997 and 1999 and will decline thereafter *if* the limits outlined in the Copenhagen Amendments are not exceeded (Montzka *et al.*, 1996).

Effective UV Dose

Latitudinal spectral UV irradiances at the Earth's surface associated with stratospheric ozone depletion were calculated using De Leeuw's (1988) atmospheric UV transfer model, which takes account of the absorption of solar radiation by ozone and aerosols, the scattering by air molecules and aerosols, the influence of clouds, and the reflection at the Earth's surface. The model includes an urban aerosol profile and a standard tropospheric ozone profile, which remain unchanged over time. The atmosphere is modelled as a series of layers, and the calculated UV radiation at ground level is reasonably consistent with the limited measurements of UV available at present (de Leeuw & Slaper, 1989; Bordewijk *et al.*, 1995).

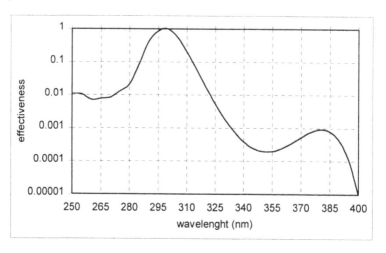

Figure 6.3: The SCUP-h Action Spectrum (de Gruijl and van der Leun (1994))

Information on the wavelength dependency (the action spectrum) is crucial for a quantitative assessment of the skin cancer risk from a given UV source. Such information cannot be determined directly for humans, and has to be inferred by other methods. A feasible alternative is to determine this action spectrum in (hairless) mice, and to convert this result to humans by correcting for optical differences between the human and mouse skin. Figure 6.3 shows the SCUP-h (skin cancer Utrecht Philadelphia-human) action spectrum for humans, used in the simulation. Although this action spectrum is determined by the induction of SCC in mice, given the parallels with BCC this action spectrum is also applicable for BCC (de Gruijl & van der Leun, 1994). Recently, results have been published on the action spectrum for melanoma skin cancer in hybrid fish (Setlow *et al.*, 1993). This melanoma action spectrum is 'flatter' than the SCUP-h action spectrum; the effectiveness of UV-A in producing melanoma is very high in this model. However, the question arises as to whether this action spectrum can be extrapolated to humans. Therefore, in this study it is assumed that the action spectrum for melanoma induction is similar to the action spectrum for the induction of NMSC.

Box 6.2:

UV Dose and Skin Cancer Rates

The annual carcinogenic effective UV dose (D (ϕ,t) in mJ/cm^2) for a given stratospheric ozone thickness O_m (ϕ,t) at latitude ϕ and at time t is calculated as the action spectrum (i.e. the weighted sum) of all spectral UV doses per month (Slaper et al., 1992):

$$D(\phi,t) = \sum_{\lambda=100}^{400}[\sum_{m=1}^{12} S_m(\phi,\lambda,t)]A(\lambda) \qquad (6.4)$$

where λ is the UV wavelength (range 100-400) (nm), S_m (ϕ,λ,t) is the total monthly spectral dose for month m and wavelength λ at latitude ϕ and at time t (mJ/cm^2), and $A(\lambda)$ is the action spectrum, i.e. the relative effective weight factor at wavelength λ.

The actual effective dose received by the skin (E (ϕ,t) in mJ/cm^2) is calculated as a fraction of the total available solar effective UV dose at the Earth's surface:

$$E(\phi,t) = \beta D(\phi,t) \qquad (6.5)$$

where β is the exposure fraction received by the most exposed parts of the skin (head and hands).

The cumulative incidence of skin cancer cases until age $l(a)$ (=$5(a-1)+2$; in the model, the population is divided into 18 age groups at 5 year intervals: 0...4, 5...9, ..., 80...84, 85+, denoted as $a=1, 2,..., 17, 18$, respectively) at time t at latitude ϕ ($Y_s(\phi,t,a)$ in cases per 100,000) involving skin cancer type s ($s=1$: BCC; $s=2$: SCC; $s=3$: MSC) is calculated as a function of the cumulative, long-term UV-B exposure (Slaper et al., 1986, 1992):

$$Y_s(\phi,t,a) = \gamma_s[\sum_{j=1}^{l(a)} E(\phi,t-l(a)+j)]^{c_s} l(a)^{d_s-c_s} \qquad (6.6)$$

where γ_s is the calibration coefficient estimated according to the skin cancer type, the sensitivity of the population, and the skin area exposed, and is estimated on the basis of the 1990 incidences described before. $\sum^{l(a)}_{j=1} E(\phi,t-l(a)+j)$ is the cumulative effective UV dose received by a person at the average age $l(a)$ in age group a (mJ/cm^2). The received effective dose before 1978 is assumed to be equal to the 1978 level (no ozone depletion); c_s, and d_s are skin cancer type-dependent model parameters describing the dose dependence and time dependence, respectively (see main text).

The annual skin cancer incidence (I_s (ϕ,t)) can now be described as (with Y_s ($\phi,t,0$) = 0):

$$I_s(\phi,t) = \sum_{a=1}^{18} N(a,t)[Y_s(\phi,t,a) - Y_s(\phi,t-1,a-1)] \qquad (6.7)$$

where $N(a,t)$ is the number of people in age group a at time t.

The total mortality rate is calculated by multiplying the annual skin cancer incidence by the lethal fractions of each skin cancer type (lethal fraction BCC=0.003; SCC=0.03; MSC=0.25).

Based on this action spectrum, the annual carcinogenic effective UV dose is calculated (see Box 6.2). Only a fraction of the total available solar effective UV dose at Earth's surface will be received by the skin. This exposure fraction depends on individual lifestyles with regard to UV exposure. For The Netherlands and Australia, this fraction (parameter β in Box 6.2) is equal to 0.02 and 0.05, respectively. The calculations of the UV doses involve the assumption of cloudless conditions, and the calculated annual carcinogenic effective UV dose ($D(\phi,t)$ in Box 6.2) therefore represents an overestimate of the actual annual dose reaching the Earth's surface. If the effects of clouds were taken into account, the UV doses would be reduced to 65 per cent of the clear-sky figures, and thus an exposure fraction β of 0.02 and 0.05, as assumed in the model, corresponds with an actual fraction of 0.03 and 0.07. This actual fraction of 0.03 corresponds to the observed average exposure experienced by indoor workers in The Netherlands (Slaper, 1987); based on studies by Hill et al. (1992) and Leach et al. (1978), for Australia this fraction is equal to the corresponding 0.07. It should be noted that other values for β do not influence the results, if the exposure factor does not change over time and/or with age (as assumed). Changes in the value of β lead to change in the estimated parameter γ_s (equation 6.6), but not to the number of additional cases of skin cancer.

Skin Cancer Incidence

SCC is most clearly related to UV exposure: these tumours occur on sun-exposed skin (face, head, neck, back of hands), and incidences in genetically and behaviourally comparable populations increase strongly in the direction of the equator (Scotto et al., 1981). Fair-skinned people who are susceptible to sunburn (poor tanners) run the highest risk of SCC. Furthermore, the SCC risk rises in relation to the total dose of sunlight received over a lifetime (Vitaliano & Urbach, 1980; Vitasa et al., 1990). At first glance the aetiology of BCC seems similar to that of SCC, e.g. fair-skinned people sensitive to sunburn run the highest risk, incidences increase toward the equator, and the condition is frequently found in the face-neck areas. However, BCC hardly occurs on the regularly sun-exposed backs of the hands, but it occurs more frequently than SCC on the trunk (10–20 per cent of the incidence in a US survey (Scotto et al., 1981), and up to 80 per cent in a cross-sectional prevalence study of people over 50 years of age in Australia (Kricker et al., 1995)). In this respect, BCC appears to resemble more the aetiology of MSC than of SCC.

Except for lentigo *maligna melanoma* (whose aetiology resembles that of SCC, but makes up approximately 10 per cent of MSC), MSCs do not mainly occur on the face and neck areas, but on a large part on the trunk (males and females) and on the lower legs (females). Retrospective studies based on questionnaires have confirmed the suspicion that intermittent (over)exposure is a risk factor, as was suggested by the localisation of the tumours over the body. However, recall bias may seriously affect this type of study (Weinstock et al., 1991). Studies on immigrants from a temperate climate to a subtropical one (e.g. from Great Britain to Australia) showed that the risk of MSC later in life rises strongly for people arriving before adolescence (the risk approaches that of people born in the subtropics) (Holman & Armstrong, 1984). Although animal experiments do not provide unique answers as to how UV radiation can cause MSC, they do make it clear that UV radiation can at least aid its development (Donawho & Kripke, 1991; Husain et al., 1991). Although MSC is less common, it is associated with a much higher rate of fatalities (about 25 per cent).

Two important variables in the assessment of the impact of increased levels of UV on skin cancer risk are c_s and d_s: skin cancer type-dependent model parameters describing the dose dependence and time dependence, respectively (see Box 6.2). The parameter c_s, which in the literature is often referred to as the biological amplification factor (BAF), refers to the fact that a 1 per cent increase in effective UV dosage leads to a c_s per cent increase in the incidence of skin cancer. The values of c_s and d_s are derived from epidemiology (see Slaper et al., 1996); c_s is equal to 1.4 for BCC and 2.5 for SCC. With respect to MSC, the relationship between UV exposure and actual incidence is somewhat ambiguous. The dose-response parameter c_s for melanoma skin cancer has been estimated as 0.6 on the basis of epidemiological data obtained at various latitudes in the USA (Environmental Protection Agency (EPA), 1987). The parameter d_s is based on epidemiological evidence of age-specific incidence and is found by fitting a least squares linear relationship between the $\ln(a)$ and $\ln(I)$, where a is the age and I the cumulative incidence. The values of d_s were estimated on the basis of age-specific incidences for three separate birth cohorts, as derived from EPA (1987) (Slaper et al., 1992). Table 6.1 sets out the values of these parameters for BCC, SCC and MSC.

Table 6.1: The Values of the Parameters c_s and d_s (Slaper et al. (1996))

Skin cancer type	c_s ($\pm$sd)	d_s ($\pm$sd)
BCC	1.4$\pm$0.4	4.9$\pm$0.6
SCC	2.5$\pm$0.7	6.6$\pm$0.4
MSC	0.6$\pm$0.3	4.7$\pm$1.0

The Netherlands

The present incidence of NMSC is not very well known in The Netherlands due to an incomplete registration (Slaper et al., 1992). Based on an extrapolation of the incidence figures of the USA presented by the third national skin cancer survey

(Scotto *et al.*, 1981), and taking into account the lower solar intensity in The Netherlands, the NMSC incidence is estimated at 18,750 cases per year (125 per 100,000). Neering & Cramer (1988) estimated the incidence at 15,000 cases yearly (100 per 100,000), and the Dutch Health Council estimated the incidence at 900 (± 300) per million per year for BCC and 160 (±20) per million for SCC (Dutch Health Council, 1994). Following Slaper *et al.* (1992), this study assumes the incidence to be 16,500 cases per year, whereby the annual incidence BCC is 100/100,000 and that of squamous cell carcinoma 10/100,000. The risk of mortality is estimated as 0.3 per cent for BCC, and 3 per cent SCC. These fractions are in accordance with the present annual mortality rate for NMSC in The Netherlands, which is 80–90 cases (6 per million). Over recent decades, the incidence of melanoma skin cancer has increased rapidly in The Netherlands. That these increases are much greater than expected is due to the changes in the size and age distribution of the population, and to ozone depletion. It is currently believed that 'sun-worshipping' holidays are a major factor explaining the rising melanoma incidence, although the evidence supporting this view is not conclusive (UNEP, 1994). The observed mortality rate has increased almost sevenfold, from 0.3 per 100,000 in 1950 to 2 per 100,000 in 1990 (an annual increase of 5–7 per cent). The present rate of 2 per 100,000 represents 300 cases per year, and in view of the present incidence of about 1200 cases (8 per 100,000) reported by Neelemans and Rampen (1990), the risk of mortality is estimated as 25 per cent.

Australia

The incidence of NMSC in Australia is the highest reported in the world (Giles *et al.*, 1988). In 1985 and 1990 surveys for the whole of Australia were carried out to detect people with recently treated lesions, approaching their doctors for medical confirmation (Giles *et al.*, 1988; Marks *et al.*, 1993). However, these studies fall short of estimating true incidences because of the large number of people with undiagnosed NMSCs. Based on these data, adjusting for the fraction of potential cases for which diagnosis was unconfirmed (multiplying the data by 1.483 (Giles *et al.*, 1988)), the estimates of the incidence of NMSC in Australia in 1990 are about 1080 cases of BCC per 100,000 per year, and 370 cases of SCC per 100,000 per year. The melanoma incidences have been collected through the National Cancer Statistics Clearing House (NCSCH) in Canberra (data from 1982–1988) and the remaining data (before 1982 and after 1988) have been obtained by contacting the individual head offices of the state cancer registries. The estimated 1990 incidence of MSC is 40/100,000, 16/100,000 and 30/100,000 for zones 1, 2 and 3 respectively. The risk of mortality for the three skin cancer types (SCC, BCC and melanoma skin cancer) are assumed to be equal to the assumed risks for the Dutch situation, namely 0.3 per cent for BCC, and 3 per cent for SCC, and 25 per cent for MSC.

OZONE DEPLETION AND SKIN CANCER RISK IN THE NETHERLANDS AND AUSTRALIA

Ozone Depletion

Netherlands

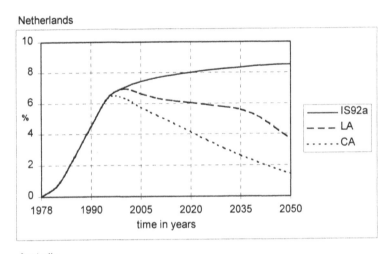

Australia

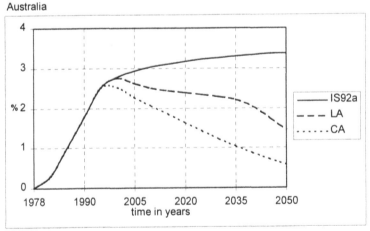

Figure 6.4: The Cumulative Depletion of the Ozone Layer due to Increase in Chlorine Levels for The Netherlands and Australia (Average of the Three Zones Considered), for the IPCC IS92a Scenario, and for the London (LA) and Copenhagen (CA) Amendments to the Montreal Protocol

Cumulative stratospheric ozone losses due to increases in chlorine levels are depicted for the three scenarios in Figure 6.4 for The Netherlands and Australia, respectively. Initial stratospheric ozone depletion in 1990 is simulated to be about 6 per cent and 3 per cent for The Netherlands and Australia respectively, which evidently, because of the calibration technique, corresponds to the negative ozone trend over the last decade as detected by the TOMS measurements (de Winter-Sorkina, 1995). The IPCC IS92a scenario shows large stratospheric ozone reductions during the coming decades. If the Copenhagen (or London) Amendments to the Montreal Protocol are fully complied with, stratospheric chlorine concentrations, and hence ozone depletion, will peak within the next few years and decrease during the next century. However, given the long half-life of chlorine radicals in the stratosphere, chlorine levels will decrease only slowly. The subsequent recovery of the ozone layer would occur by replenishment, and, in the case of the Copenhagen Amendments, the 1970 levels would be recovered by the middle of the next century.

Skin Cancer Incidence and Mortality Rates

The skin cancer incidence projections presented in this section are all expressed in terms of excess rates of skin cancer (or deaths due to skin cancer), defined as the extra number of skin cancer cases or deaths due to skin cancer per year expected as a consequence of stratospheric ozone depletion. These are calculated by subtracting the baseline estimates, i.e. the incidences occurring when no depletion of ozone takes place, from the total incidences obtained according to the various scenarios.

Because older people build up a high cumulative UV dose during their lives, skin cancer occurs primarily among the elderly. In an ageing population the same level of UV exposure would lead to higher incidence than in a 'younger' population. Over recent decades the Dutch and Australian population has been ageing, and this trend will continue in the foreseeable future. To distinguish the effects of ozone depletion on skin cancer rates from the combined effects of ozone depletion and a changing age distribution of the population, the model is run for a standardised population (the 1990 age distribution in The Netherlands and Australia; the constant 1990 population) and a population growth scenario. Historical population data are used over the period 1950–1990.

Figure 6.5 shows the results for the excess rates of skin cancer for the three halocarbon scenarios. For the Dutch constant 1990 population scenario, the excess rate of NMSC increases from about 1 per 100,000 in 1990 to ~15 per 100,000 in 2050 for the IS92a scenario. For the London and Copenhagen Amendments to the Montreal Protocol, the excess rate increases up to ~11 and 7 cases per 100,000 by the year 2050, respectively (~12 and 8 times 1990 levels). The excess rates for MSC in 2050 are about 0.2, 0.2, and 0.1 cases per 100,000 for the IS92a and the London and Copenhagen Amendments, respectively (16, 11, and 8 times higher than the 1990 levels).

Trends in Australia exhibit the same patterns. Although the excess numbers of skin cancers per 100,000 population are higher (because of the higher initial

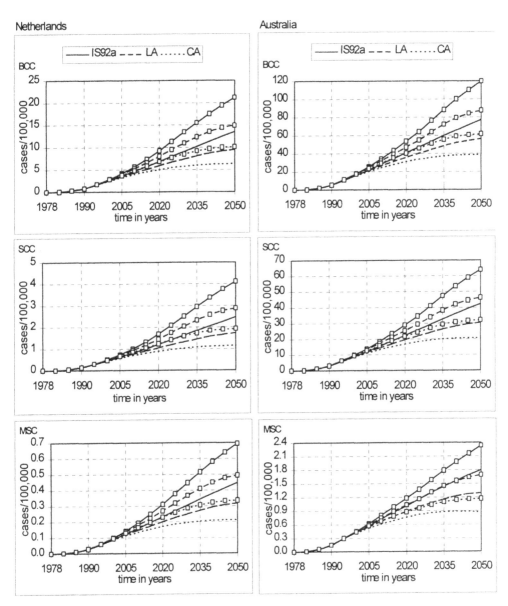

Figure 6.5: Excess Incidence Rates of BCC, SCC and MSC for The Netherlands and Australia for the IPCC IS92a Scenario, and for the London (LA) and Copenhagen (CA) Amendments to the Montreal Protocol: Results Are Generated with a Constant 1990 Age Distribution (Constant Population Scenario), and a Dynamic Population (Population Growth Scenario; Marked by □)

cancer incidences), the relative increases compared to 1990 are lower for the Australians. This is a result of the more rapid ozone depletion at higher latitudes and the age distribution of the population at risk (the Dutch population is relatively 'older' than the Australian population), which, when comparing The Netherlands with Australia, is not 'compensated' by the higher increase in effective UV-dose per per cent ozone depletion at lower latitudes. For NMSC the increases in cancer rates are 119, 86, and 60 cases/100,000 (~14, ~10, and ~7 times the 1990 rates) for the three scenarios, respectively. For MSC the excess incidence is 1.8, 1.3, and 0.9 ~12, ~9, and ~6 times the 1990 rates), with the largest increases in the Australian zones 1 and 2.

Figure 6.5 also shows the skin cancer incidences obtained by adopting a population growth scenario for The Netherlands (Central Bureau of Statistics (CBS), 1995) and Australia (Australian Bureau of Statistics (ABS), 1994) over the period 1990–2050. For both countries the mid-estimate (projection B for Australia as described in ABS (1994)) has been used; for The Netherlands, the population is assumed to increase to ~17 million people in the year 2030, with population numbers decreasing thereafter. The projection used for Australia leads to a population of about 25 million people in 2050. In general, the excess rates of non-melanoma carcinoma and MSC are now about 50-60 per cent higher compared to the estimates for the constant 1990 population case due to ageing of the population. Table 6.2 summarises the results. Figure 6.6 shows the excess mortality rates for the Copenhagen Amendments to the Montreal Protocol for the population growth scenarios. It appears that *excess* mortality rates in The Netherlands are mainly due to MSC, whereas in Australia, as a consequence of the larger number of SCC cases relative to MSC (compared to The Netherlands), excess mortality due to SCC dominates. However, both for The Netherlands and Australia, MSC remains the skin cancer responsible for most of the mortality.

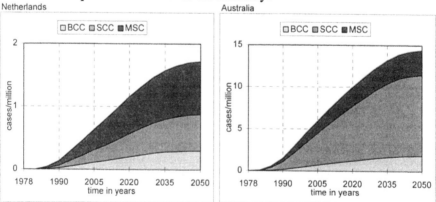

Figure 6.6: Excess Mortality Rates of BCC, SCC and MSC for The Netherlands and Australia for the Copenhagen Amendments (CA) to the Montreal Protocol: Results Are Generated with the Population Growth Scenario

Table 6.2: Summary of the Excess Incidence and Mortality Rates (per 100,000) of Simulation Runs for the Constant 1990 Population (cp) and the Population Growth (pg) Scenario for The Netherlands (Net) and Australia (Aus) in the Year 2050

Scenario	Net						Aus					
	IS92a		LA		CA		IS92a		LA		CA	
	cp	pg	cp	pg	cp	pg	cp	pg	cp	pg	cp	pg
BCC	13	21	9	15	6	10	77	120	56	88	39	61
SCC	2	4	2	3	1	2	42	64	30	46	21	32
MSC	0.4	0.7	0.3	0.5	0.2	0.3	1.8	2.3	1.3	1.7	0.9	1.2
Total mortality	0.2	0.4	0.2	0.3	0.1	0.2	2.0	2.9	1.4	2.1	1.0	1.4

A comparison of the excess skin cancer incidence in 1990 for The Netherlands and Australia with the actual incidences as used in this study shows that the modelled skin cancer incidence fraction due to present stratospheric ozone depletion above the two countries does not exceed 1 per cent. This rather small increase is in accordance with the time delay of several decades between the time of increased UV-B levels and the development of skin cancer.

UNCERTAINTIES

To examine the extent to which variations in various main model parameters, representing the processes of ozone depletion, effective UV-B radiation, and skin cancer development, affect the excess rates of the incidence of skin cancer, an uncertainty analysis is performed in this section. The analysis is conducted using the Copenhagen Amendments and the population growth scenarios and is summarised in Table 6.3. The reference scenario is the Copenhagen Amendments scenario with 'central values' of k, β, and c_S (the excess incidence rates appeared to be rather insensitive to changes in the parameter d_S).

Table 6.3: Summary of the Results of the Sensitivity Analysis: Excess Incidence Rates per 100,000 (Percentage Change Compared to Reference) in the Year 2050, for the Copenhagen Amendments and the Population Growth Scenario

	Ref	Max. k	Min. k	Min. β	Max. β	Min. c_S	Max. c_S
Netherlands							
BCC	10	6 (-45)	15 (50)	5 (-55)	17 (65)	4 (-58)	16 (59)
SCC	2	1 (-46)	3 (51)	1 (-73)	5 (133)	1 (-57)	3 (60)
MSC	0.3	0.2 (-45)	0.5 (48)	0.2 (-34)	0.4 (28)	0 (-100)	0.7 (102)
Australia							
BCC	61	11 (-81)	116 (90)	28 (-55)	101 (65)	26 (-58)	97 (59)
SCC	32	2 (-94)	67 (107)	9 (-73)	75 (134)	14 (-57)	51 (58)
MSC	1.2	-0.2 (-117)	2.8 (141)	0.8 (-34)	1.5 (28)	0 (-100)	2.4 (101)

Ozone Depletion Rates

The sensitivity of changes in future ozone depletion rates on the excess skin cancer incidence rate can be examined by varying the coefficient k in equation 6.1 between its maximum and its minimum value (the minimum and maximum k scenarios refer to a k factor of minus or plus twice the standard deviation). For The Netherlands the calculated range of the maximum cumulative ozone depletion is between ~5 and 8 per cent around the year 2000 (~5 per cent for the reference case); for Australia it lies between ~1 and 4 per cent (reference case ~3 per cent), with the largest ozone depletion in the highest latitude zone (zone 3). The simulated excess incidence rates for low values of k (the lower the value of k, the larger the ozone depletion) are about 50–140 per cent higher than those associated with the reference case, with a significant difference between the excess rates of NMSC and MSC for Australia; no significant difference occurs for The Netherlands. The simulated changes compared to the reference suggest that MSC responds more rapidly to higher, peak UV doses. The relatively larger uncertainty range surrounding the percentage ozone depletion per ppbv increase in chlorine concentration for Australia compared to The Netherlands is reflected in the difference in the changes of the excess skin cancer rates.

Biological Amplification Factor (BAF)

Estimates for the dose-response relationship between UV radiation and skin cancer incidence are based on various action spectra, and may vary widely. Generally, for BCC the estimates of the c_s (BAF) in equation 6.6 from various studies lie between 1.1 and 2.6. For SCC and melanoma there is a much greater variation in estimates. Those for SCC are between about 1.0 and 4.0; for melanoma these estimates vary between 0.3 and 3.0. Many assumptions surround the estimates of the c_s, such as the correctness of the action spectrum being used, and the cancer incidences have been measured accurately. However, as none of them is likely to be correct, any estimates made of c_s so far are likely to be inaccurate (WHO, 1994).

Table 6.3 gives the influence of changes in the BAF on the excess incidence rates as simulated with the model. As is the case with k, the minimum and maximum c_s scenarios refer to a c_s factor of minus or plus twice the standard deviation, as presented in Table 6.1, respectively. The results in Table 6.3 reflect the high uncertainty surrounding the UV dependency of MSC. The ranges surrounding the estimates of BCC and SCC are about the same.

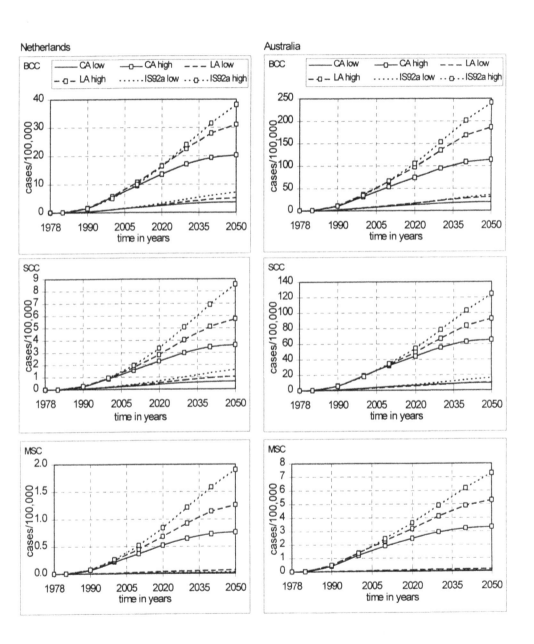

Figure 6.7: Uncertainty in the Excess Skin Cancer Rates for the IPCC IS92a Scenario, and for the London (LA) and Copenhagen (CA) Amendments to the Montreal Protocol, with a Population Growth Scenario

Changing exposure habits

To study the influence of exposure habits on the skin cancer risk estimates, the exposure fraction parameter β is varied by ± 50 per cent over the period 1990–2050. The results, presented in Table 6.3, show that a change in exposure habits would mainly affect the excess incidence rate of non-melanoma carcinoma: an increase of ~65–135 per cent for a high exposure, and a decrease of ~55–75 per cent for a low exposure, vis-à-vis the reference case. For MSC the change compared to the reference scenario is about ± 30 per cent. This difference between NMSC and MSC can be explained by the higher c_s factor for NMSC, which implies that the same change in effective UV dose would have a greater effect on NMSC incidence than on MSC incidence.

Total Uncertainty

The previous sections have described the individual influence of changes in model parameters on the excess skin cancer rates. This section explores the total uncertainty surrounding the estimates of the excess skin cancer rate, associated with uncertainties of the model parameters combined (including parameter d_s). The total uncertainty of the simulated excess rates is calculated in a Monte Carlo procedure (using a number of 100 samples; selecting random variables from the uniform probability distribution (minimum and maximum values as described above) for the model parameters). Figure 6.7 presents the uncertainty range (the 95 per cent 'confidence band' range between the 2.5 per cent and 97.5 per cent percentiles) for the excess skin cancer incidence for the three skin cancers considered.

The simulated excess incidence for the reference case (all model parameters set to their central value, as presented in the previous sections), does not significantly differ from the mean values generated here, and is less than 2 per cent. High uncertainties surround the estimate of the impact of ozone depletion on skin cancer rates, reflected by the large overlap between the uncertainty range of the three scenarios. Note that the uncertainties in the estimation of the chlorine concentration in the atmosphere, which would further influence the uncertainty range, are not taken into account in this analysis (see den Elzen, 1993).

DISCUSSION AND CONCLUSIONS

First of all, the uncertainties surrounding the trends in total stratospheric ozone and chlorine concentration during the period 1978–1991, on which the relationship between chlorine concentration and ozone depletion (coefficient k) is calibrated, will translate into major uncertainties in future ozone depletion and the consequent skin cancer incidences. For tropical regions, trends in total ozone depletion are less significant (Stolarski *et al.*, 1991), leading to larger uncertainties in the estimation of future ozone depletion for the Australian continent, compared to The Netherlands. For example, a high estimate of the increase in the rate of ozone depletion would increase the excess skin cancer incidence by 50 per cent for The Netherlands, and by 140 per cent for Australia. Using the higher ozone trends in the most recent years (probably influenced by the volcanic eruption of Mount Pinatubo in 1991 (de Winter-Sorkina, 1995)) would increase the estimates of additional skin cancer rates. Furthermore, using a smaller trend in chlorine concentration than the value used (e.g. comparing the value of 0.11 ppbv/year used in this study with 0.085 ppbv/year as used in Slaper *et al.* (1996)), would lead to higher estimates of *k*, and consequently higher estimates of the excess skin cancer rates.

Second, although there is a large body of data, both experimental and epidemiological, that confirms a causal relationship between accumulated UV dose and SCC (e.g. Forbes *et al.*, 1978; Vitaliano & Urbach, 1980), the UV dose dependencies of BCC and MSC (except for lentigo *maligna melanoma*) are less certain. Model calculations show that the incidence of MSC is possibly more affected by a high UV-B peak dose than by moderate, increased UV-B levels over a longer period. Therefore, the future peak in ozone depletion, expected around the year 2000 if the Copenhagen Amendments are adhered to, could be of major importance for the coming skin cancer incidences in the next century. Furthermore, MSC has been induced in fish, UV-A being highly effective in inducing the melanomas (Setlow *et al.*, 1993). If this holds for humans too, then MSC would not react appreciably to ozone depletion (Slaper *et al.*, 1996). However, this would hardly affect the estimates of the excess skin cancer rates due to ozone depletion, as MSC only contributes less than 5 per cent to the total incidence rates. Excess mortality rates, on the other hand, would decrease by ~50 per cent and ~20 per cent for The Netherlands and Australia, respectively.

Another important aspect is that skin cancer rates are very sensitive in respect of lifestyle (i.e sun exposure habits). Changing lifestyles (e.g. 'sun worshipping') at present contribute greatly to the increases in the incidence of skin cancer, which already impose a steadily increasing strain on the medical care system. This has been identified as a serious public health problem in several Western countries, and campaigns have been launched to curb excessive exposure to the sun (e.g. the 'Watch your skin' campaign by The Netherlands Cancer Foundation). In Australia, more and more people stay out of the sun during the midday hours as a reaction to the alarming number of skin cancer occurrences in that country. Another factor contributing to a steady increase in the number of skin cancers is the ageing of the population, which may cause a 50–60 per cent increase in the overall incidence.

The results presented clearly show the delay mechanisms in the effect of ozone depletion on skin cancer rates. Full compliance with the Copenhagen Amendments would lead to a peak in the tropospheric chlorine concentration around 1995, a peak in stratospheric chlorine concentration and ozone depletion around 2000, and to a peak in skin cancer by about 2050 (50 years after the peak in ozone depletion). The latter delay is mainly due to the fact that skin cancer incidences depend on the cumulative UV-B exposure suffered by an individual. In view of the several delay mechanisms and, additionally, the ageing of the population, future increases in skin cancer incidence are likely to occur.

Chapter 7

DISCUSSION AND CONCLUSIONS

INTRODUCTION

The scientific and policy community were slow to recognise the potential importance and scope of human health impacts of global atmospheric changes, and only a small amount of scientific literature on the subject has been generated to date. Given the many uncertainties in the health impact assessment and the complexity of the processes involved, many assessments have been qualitative, or semi-quantitative. Although epidemiology is the basic quantitative science of public health, only for a minority of the expected impacts of climate change and ozone depletion, such as mortality due to thermal stress, is an extension of the standard epidemiological risk assessment possible. For other impacts, (new) modelling techniques are required. However, the construction of integrated models for the health impact assessment of global atmospheric changes is still at a relatively early stage of development. Therefore, the 'first generation' models presented in this study are meant to increase our insights into the underlying processes of climate change, ozone depletion and human health, and intended to stimulate and contribute to the ongoing discussion of the development of methods in the analysis of the interactions between atmospheric changes, ecosystems and human health.

MAJOR FINDINGS

Eco-Epidemiological Modelling Approach

Conventional epidemiological methods are often poorly suited to studying the health impacts of climate change and ozone depletion. Compared to conventional (environmental) epidemiology, three major polarities characterise this new research domain: (i) spatial scale, i.e. regional/global versus local impacts; (ii) temporal scale, i.e. future versus present health risks; and (iii) level of complexity, i.e. complex eco-epidemiological processes versus straightforward cause-effect relationships.

Recognising the problem that mainstream epidemiological methods are often not well adapted to the analysis of disease causation that involves complex systems influenced by human interventions or 'simpler' processes which will take place in the (distant) future, a different way of looking at these problems is required. My view, as discussed in the previous chapters, has been an eco-epidemiological modelling one: using integrated mathematical models to provide a better quantitative understanding of the dynamics underlying the health impacts of climate change and ozone depletion (and the sensitivities and uncertainties which surround these impacts), which may arise within an ecological context, with maximum reference to existing epidemiological knowledge of disease causation. Within MIASMA, integrated assessment models have been developed to analyse the impact of climate change on vector-borne diseases and mortality changes related to thermal stress, and the effect of stratospheric ozone depletion on skin cancer rates.

The effect of climatic changes on malaria, schistosomiasis, and dengue is probably the clearest example of a health impact with complex climate-related, ecologically based dynamics. However, besides incorporating ecological components (e.g. changes in vector distribution are linked to changes in vegetation patterns), the system-dynamic models developed to assess climate impacts on vector-borne diseases are to a large extent based on infectious disease epidemiology. Acknowledging the fact that the systems approach used to model the impact of climatic changes on vector-borne diseases was not suitable to simulate adaptive processes, as a case study a complex adaptive systems approach was used to simulate the development of resistance among the malaria parasite and mosquito populations. The assessment of the mortality changes related to thermal stress as a result of a change in ambient temperature and the effects of increased UV-B radiation on skin cancer rates has a clear epidemiological basis, in that most of the input parameters are derived from epidemiological studies.

Climate Change and Vector-Borne Diseases

Simulations using the vector-borne diseases model, employing climate change scenarios from three GCMs combined with the EP index, show an increase of the populations at risk of malaria, dengue and schistosomiasis. Assuming that population in the developing world will increase up to a total of ~8.6 billion by the year 2050, the *additional* number of people at risk due to anthropogenic climate change may increase up to about 720 million, 40 million, and 195 million people for malaria, schistosomiasis, and dengue, respectively. There would be an increase of the risk of local transmission of the three vector-borne diseases in developed countries, associated with imported cases of the disease. However, given the fact that effective control measures are economically feasible in these countries, it is not to be expected that human-induced climate changes would lead to a return of a state of endemicity in these areas.

For the three climate change scenarios used an increase in the EP of the malarial and dengue parasite in the presently vulnerable regions in the (sub)tropics is to be anticipated. Although surrounded by large uncertainties, on aggregate this varies between ~10 and 30 per cent for malaria and between ~30 and 50 per cent for dengue, towards the year 2050. As a consequence of the relatively low optimum transmission temperature given a certain snail density, the transmission potential of schistosomiasis may decrease by between ~10 and 20 per cent in the present potential areas. However, in the current highly endemic areas, the prevalence of infection is persistently high, and will probably only be marginally affected by these climate-induced changes. In view of their high potential receptivity and the immunological naivety of the population, the highest risks for the intensifying of transmission of malaria, dengue and schistosomiasis reside in regions of hitherto no or low endemicity on the altitudinal and latitudinal fringes of disease transmission. An aspect of particular importance is the increase EP at higher altitudes within endemic areas such as the Eastern Highlands of Africa, the Andes region in South America, and the western mountainous region of China, illustrated by the simulated intensification of malaria transmission at high altitudes in Zimbabwe and the increase of dengue EP in several cities that are presently only at a low risk.

Modelling Malaria as a Complex Adaptive System

The systems-based approach used in the impact assessment of climate change on malaria risk does not consider the ability of the (malaria) systems to adapt to changes. To simulate the development of resistance of the malaria mosquito and parasite to pesticides and drugs, respectively, a genetic algorithm was included in the systems-based model. This evolutionary modelling approach fits well within the present qualitative notion of the importance of evolutionary principles for infectious diseases. Modelling malaria as a complex adaptive system illustrates that, although *adequate* use of insecticides and drugs may reduce the occurrence of malaria, climate change may intensify the growing problem of the development of resistance to control programmes. However, in the highly endemic regions the use of insecticides and drugs may lead to increased incidence due to enhanced resistance development. Migration of (susceptible) mosquitoes and parasites may significantly retard the development of resistance. Although a great deal of empirical research is needed to improve this modelling approach, the development of integrated assessment models based on the evolutionary and local dynamics of ecological systems may be essential in assessing future developments of these complex adaptive systems. Together with the expected relevance of, inter alia, land use changes and patterns of human migration, the next step in developing an integrated assessment tool for malaria is the inclusion of spatial characteristics, based on cellular automata, for example.

Climate Change, Thermal Stress and Mortality Changes

Data from several epidemiological studies on the relationship between temperature and mortality suggest that total and respiratory mortality are more sensitive to a 1°C increase at high temperatures (i.e. temperatures above the comfort range of about 16°C–23°C) than to a 1°C increase in the 'cold' temperature range (temperatures below the comfort range), in contrast to CVDs. On aggregate, a 1°C increase of monthly mean temperature may increase total, respiratory, and cardiovascular mortality by 1.4 per cent, 10.4 per cent, and 1.6 per cent, respectively, if the temperature exceeds the comfort range. Below this comfort range, a 1°C increase may decrease mortality rates by 1.0 per cent, 3.8 per cent, and 4.1 per cent, for total, respiratory diseases and CVDs, respectively. However, for total and respiratory mortality the estimates are based on only a few studies, and must therefore be interpreted with caution. Simulations show that a globally averaged temperature increase of ~1.2°C (to be expected somewhere in the time frame of 2040–2100) could, in areas with relatively colder climates result in a reduction of the disease burden due to less excess winter cardiovascular mortality, especially in elderly people. Based on current mortality levels, this decrease could be about 50 people per 100,000 population, whereas the increase in cardiovascular mortality in warmer climates may be about 3/100,000. However, as is the case with the impact of climate change on vector-borne diseases, different populations, with different characteristics and circumstances, respond differently to climate change-induced thermal stress. The changes in temperature-related mortality will occur against a background of changes in levels of incidence of cardiovascular and respiratory diseases through changes in prevailing lifestyles (e.g. diet, smoking behaviour), age distributions, levels of air pollution, etc. Developing countries are likely to be more vulnerable due to rapid urbanisation, few social resources and a low pre-existing health status.

Ozone Depletion and Skin Cancer Rates

The study on ozone depletion and skin cancer clearly shows the delay mechanisms in the effect of ozone depletion on skin cancer rates. Full compliance with the Copenhagen Amendments would lead to a peak in the tropospheric chlorine concentration around 1995, a peak in stratospheric chlorine concentration and ozone depletion around 2000, and a peak in skin cancer by about 2050 (50 years after the peak in ozone depletion). The latter delay is mainly due to the fact that skin cancer incidences depend on the cumulative UV-B exposure suffered by an individual. Excess NMSC rates may increase up to about 7 and 60 cases per 100,000 in the year 2050, for The Netherlands and Australia respectively, assuming a constant 1990 population (compared with ~120 and ~1540 cases/100,000, provided no (further) depletion of the ozone layer were to occur). For melanoma, the excess rates are ~0.1 and ~1 cases per 100,000 (relative to ~8 and ~85 cases/100,000). However, although there is a large body of data, both experimental and epidemiological, that confirms a causal relationship between accumulated UV dose and SCC, the UV dose dependencies of BCC and MSC (except for lentigo *maligna melanoma*) are less certain. Consequently, the impacts of ozone depletion on these incidences can only be assessed with some reservation, and many uncertainties surround this problem. For example, a high estimate of the increase in the rate of ozone depletion could increase the excess skin cancer incidence by 50 per cent for The Netherlands, and by 140 per cent for Australia. Another important factor is the ageing of the population, which may cause a 50–60 per cent increase in the overall incidence towards the year 2050. Despite the uncertainties, in view of the numerous delay mechanisms and, additionally, the ageing of the population, future increases in skin cancer incidence are likely to occur, although a reduction in the exposure of human skin to UV-B radiation would largely reduce the excess incidence rates.

Synthesis

The issues addressed in this study demonstrate that global climate change and the depletion of the ozone layer are likely to influence human health in various ways. Table 7.1 gives an indication of the relative global importance of the health impacts discussed in this study (for more details on the assumptions made in the calculations, see also the relevant chapters). Although some effects may be beneficial (e.g. in areas with relatively colder climates, an increase in ambient temperature could result in a decrease of cardiovascular mortality), most are

Table 7.1: Indication of the Relative Importance on a Global Scale of the Health Impacts of Climate Change[a] and Ozone Depletion in Terms of Mortality Changes[b]: Columns 3 and 5 Show Excess Numbers; the Background Estimates are in Parentheses (i.e. Numbers due to Population Growth Only)

Health effect	Population at risk (millions) Present	2050[g]	Total mortality Present	2050[g]	Increased risk (per cent)[h]
Malaria[c]	2400	100-700 (5200)	2-3 million	1.5-2 million (4 million)	40-50
Schistosomiasis[d]	600	10-40 (1300)	40,000	-15,000--10,000 (90,000)	-20--10
Dengue[d]	1800	0-200 (3900)	25,000	20,000-30,000 (50,000)	40-65
Skin cancer[e]	600	- (1000)	16,000	1000-2000 (26,000)	5-10
Thermal stress[f] (CVDs)	2900	- (5000)	14 million	-600,000 (24 million)	-5

a: Global mean temperature increase of ~1.2°C.

b: World population growth according to the IS92a scenario (in 2050 ~10 billion people).

c: Based on model calculations described in Chapter 3 (case fatality rate: 4 per cent in children below the age of 5, others 1 per cent).

d: Although the relation between transmission potential and mortality is complex, a rough estimate is calculated as the baseline mortality/million in population at risk (not shown in table) multiplied by the change in EP (Table 3.4), multiplied by the population at risk in 2050. Present mortality based on Murray & Lopez (1994) and WHO (1995c).

e: Using mortality changes for The Netherlands (constant population) extrapolated to rest of world (Table 6.2); assuming a world-wide sensitive population of ~10 per cent.

f: Using a constant population distribution and assuming that only the urban population is at risk of thermal stress (~50 per cent of the world population (World Bank, 1993)). Present mortality based on Murray & Lopez (1994).

g: The range surrounding the estimates is based on the different climate and ozone depletion scenarios, as discussed in the previous chapters, with central parameter inputs.

h: Excess mortality relative to background estimates.

expected to be adverse (e.g. an increase in skin cancer rates and vector-borne disease incidence is to be expected). Some impacts would occur via direct mechanisms (e.g. UV-related skin cancer, morbidity and mortality related to thermal stress); others would occur through indirect mechanisms (e.g. transmission of vector-borne diseases). The approach to be used to assess these health risks depends to a large extent on the problem being studied. In general, more straightforward health impacts, such as the impact of changes in ambient temperature on mortality, can be studied using the simple extrapolation of dose-response relationships. However, more complex processes, such as climate effects on malaria transmission (whether or not influenced by resistance development), require a (complex) systems-based approach.

Although Table 7.1 describes the health impacts as if they were to occur in isolation, it is important to emphasise the fact that there would be many cross-linkages between these health impacts of climate change and ozone depletion and changes in other social, biological and ecological circumstances. For example, there appears to be a widespread increase in the tempo of new and emerging infectious diseases (Levins *et al.*, 1994), which probably reflects a combination of demographic and environmental (climate) changes, in addition to increases in drug and pesticide resistance (Morse, 1991). Rates of disease and deaths due to cigarette smoking are likely to increase in many countries (Peto *et al.*, 1994), and rates of chronic non-infectious disease (especially heart disease, diabetes and certain cancers) in rapidly developing countries are increasing (World Bank, 1993). This complex balance sheet makes it difficult to estimate the net impact of climate change and ozone depletion on human population health.

However, the estimates in Table 7.1, despite their imprecision, show that the largest changes in absolute mortality numbers are (not surprisingly) to be expected to result from diseases which already contribute a lot to mortality numbers world-wide (e.g. malaria and CVDs). Of the health impacts studied, the impact on vector-borne diseases (especially malaria and dengue) would probably predominate. The excess mortality due to skin cancer as a result of stratospheric ozone depletion is negligible on a global scale compared to the excess malaria mortality to be expected. However, these global estimates mask regional differences, and for several countries, such as The Netherlands and Australia, the estimated increase in skin cancer rates may be a problem of concern.

Different populations, with varying levels of natural, technical and social resources, differ in their vulnerability to the health impacts. Furthermore, whether all the potential public health implications of climate change and ozone depletion will be realised depends on the degree of mitigation and adaptation which will be feasible, acceptable and economically affordable. For example, a non-malarious country with a well functioning public health system may resist the threat of malaria without even increasing health sector expenditures. On the other hand, as mitigation and adaptation options for countries, at least in part, depend on their economies, development trajectories which already compromise 'sustainability', as is the case in most poor developing countries, will leave the burden of disease due to atmospheric changes either unmitigated or only mitigated at high cost to the economy.

FUTURE RESEARCH LINES

Planning for the protection of human health from the potential impacts of global climate change and increasing UV radiation as a result of ozone depletion requires a greatly improved understanding of the disease-inducing mechanisms involved, possible synergetic effects, and the vulnerability of populations. An important aspect would be the development of theoretical and conceptual methods for the assessment of the health impact of global environmental changes. Current mainstream epidemiological research methods are not always suited to adequately address health impacts that arise within a systems-based context, i.e. a context in which the ecological and other biophysical processes display non-linear and feedback-dependent relationships. Consequently, new scientific techniques will be needed, including a substantial reliance on mathematical models.

The development of multi-disciplinary, integrated assessment models, such as those discussed in this study, would thus need to be continued. Ideally, such models should incorporate: (i) projections of global environmental changes; (ii) extrapolations of the consequent human health impacts from existing epidemiological data (or other analogous historical experience); (iii) other relevant biological theory; and (iv) models of societal response. Although the models presented in this study are still at a relatively early stage of development, lack local resolving power, and have yet to be fully validated, these 'first generation models' demonstrate the basic feasibility of quantifying the health impacts of global atmospheric changes. Furthermore, most of the current models for impact assessment focus primarily on the macro dynamics of the processes of concern. To improve our understanding of the complex dynamics of the systems considered, such as vector-borne infectious disease transmission, a subsequent step would be the development of assessment models which include evolutionary, spatial, and local dynamics of ecological systems. The complex adaptive system used to simulate the development of malaria resistance, as discussed in Chapter 4, is an example of a model that simulates adaptive processes. However, inclusion of spatial dynamics (e.g. migration, land use changes) would be another essential step in assessing the impact of (anti)malarial policies.

Much of the modelling of human health impacts will require the superimposition of data on, for example, disease incidence, vector populations, demographics, and climate, with linkage to specific geographical locations. Geographic information systems, which are computerised mapping systems, can assist in the organisation and analysis of climate, environment and disease data (Glass *et al.*, 1993). Remotely sensed data from satellite imagery would be especially useful in areas where data on population distribution, land use patterns, or transportation patterns are unavailable (Glass *et al.*, 1993; Patz & Balbus, 1996). For example, satellite-generated habitat maps have been used to project the regional risk of African sleeping sickness, which is carried by tsetse flies (Rogers & Randolph, 1991). Data derived from such systems should be integrated at an early stage in the development of integrated mathematical models.

It will be essential to integrate modelling experiments, as discussed in this study, with the monitoring of environmental health indicators. For example, in the sensitive areas bordering endemic regions, identified in Chapters 3 and 4, enhanced surveillance and response would be an essential step in recognising and thereby mitigating the emergence of the vector-borne diseases considered, whether caused by climatic changes, resistance development, or other factors. Attention should be directed towards sentinel diagnostic centres in these sensitive areas, not only to provide an early warning system, but also to improve our knowledge of climate-related diseases and to facilitate the improvement of current models. To enhance our understanding of the effects of ozone depletion on skin cancer rates (but also on cataracts and immune suppression), it is important to improve the monitoring of ozone trends and UV ground-level radiation, as well as skin cancer incidence over a range of latitudes. Likewise, this would improve our risk assessment and probably reduce the large uncertainties surrounding the estimates presented in Chapter 6.

Of course, continued research would be necessary to improve our understanding of global atmospheric change-human health relationships. Examples of such research questions, related to the subjects described in this study, include: "What empirical evidence is there of indirect climatic influences on changes in vector-borne diseases?" "What is the role of acclimatisation (whether natural or technical) in the assessment of the balance between heat-related and cold-related deaths in different geographic and population settings?" "What is the precise relationship between UV radiation and MSC, i.e. what wavelengths are most effective and at which stage of tumour development?" Other important topics include a further analysis of the impact of increased UV-B levels on immune functioning and disease incidence, determination of the dose-response relationship between UV-B and cataracts, and an assessment of how levels of malnourishment, immunosuppression and infectious diseases interact. Empirical epidemiological studies of recent climate/health relationships in areas where regional climate change has occurred (for whatever reason), as a partial analogue of future climate change impacts, could assist in revealing some of the relationships mentioned above.

EPILOGUE

At the end of a book like this, one may ask the question: "What has been the value of the models presented?" Like a telescope or a microscope, simulation models are tools – tools to see new things or to see things more clearly (Rotmans & De Vries, 1997). The model-supported analyses – described in the previous chapters – add at least two elements to the more speculative statements of the future of the globe. First, the quantification of future trends is based on a variety of 'stylised facts' and insights which are at the core of the relevant scientific disciplines. Although these insights do not ensure that future surprises do not occur, they nevertheless complement less quantitative analyses of the future and gives environmental policy a clearer description of the anticipated future health impacts. Second, the models consider future trends from a more integrative perspective than usually is the case. This provides a consistency (usually absent in more narrowly based analysis), increasing the plausibility of the resulting projections of the future. Furthermore, several of the models presented have quite novel elements too.

However, many uncertainties remain regarding the health impacts of climate change and ozone depletion. The assessment of future health impacts of these atmospheric changes remains therefore a major challenge for the scientific community. I hope that the simulation models and methods discussed in this book will be seen as small steps forward toward an improvement of our knowledge of the health impacts of climate change and ozone depletion.

REFERENCES

ABS (1994). *Projections of the populations of Australia states and territories 1993 to 2041.* Catalogue no. 3222.0. Commonwealth of Australia, Canberra.

Alderson, M.R. (1985). Season and mortality. *Health Trends,* 17, 87-96.

Anderson, R.M. & May, R.M. (1979). Prevalence of schistosome infections within molluscan populations: observed patterns and theoretical predictions. *Parasitology,* 79, 63-94.

Anderson, R.M. & May, R.M. (1985). Helminth infections of humans: mathematical models, population dynamics, and control. *Advances in Parasitology,* 24, 1-101.

Anderson, R.M. & May, R.M. (1991). *Infectious diseases of humans: dynamics and control.* Oxford University Press, New York.

Anderson, R.M., Mercer, J.G., Wilson, R.A. & Carter, N.P. (1982). Transmission of *Schistosoma mansoni* from man to snail: experimental studies of miracidial survival and infectivity in relation to larval age, water temperature, host size and host age. *Parasitology,* 85, 339-360.

Anderson, T.W. & Le Richie, W.H. (1970). Cold weather and myocardial infarction. *Lancet,* i, 292-296.

Anonymous (1992). Dengue activity in Puerto Rico, 1991. *Dengue Surveillance Summaries,* 64, 1-4.

Aron, J.A. & May, R.M. (1982). The population dynamics of malaria. In: Anderson, R.M. (ed). *The population dynamics of infectious diseases: theory and applications.* Chapman and Hall, London, 139-179.

Aron, J.L. & Silverman, B.A. (1994). Models and public health applications. In: Scott, M.E & Smith, G. (eds). *Parasitic and infectious diseases: epidemiology and ecology.* Academic Press, San Diego, CA, 73-81.

Arthur, W.B. (1990). Positive feedbacks in the economy. *Scientific American,* February, 92-99.

Babayev, A.B. (1986). Some aspects of man's acclimatisation to hot climates. In: *Climate and human health: WHO/UNEP/WMO international symposium.* Volume 2, Leningrad, 125.

Bailey, N.T.J. (1982). *The biomathematics of malaria.* Griffin, London.

Bainton, D., Moore, F. & Sweetnam, P. (1977). Temperature and deaths from ischaemic heart disease. *British Journal of Preventive and Social Medicine,* 31, 49-53.

Baker, C.B., Eischeid, J.K., Karl, T.R. & Diaz, H.F. (1994). *The quality control of long-term climatological data using objective data analysis.* Pre-prints of AMS ninth conference on applied climatology, Dallas, TX, 15-20 January, 1995.

Baker-Blocker, A. (1982). Winter weather and cardiovascular mortality in Minneapolis-St Paul. *American Journal of Public Health,* 73(3), 261-265.

Baptista, D.F., Vasconcelos, M.C. & Schall, V.T. (1989). Study of *Biomphalaria tenagophila* (orbigny, 1835) and schistosomiasis transmission in 'Alto da Boa Vista', Rio de Janeiro. *Men. Inst. Oswaldo Cruz*, 84, 325-332.

Bekessy, A., Molineaux, L. & Storey, J. (1976). Estimation of incidence and recovery rates of *Plasmodium falciparum* from longitudinal data. *Bulletin of the World Health Organisation*, 54, 685-693.

Beneson, A.S. (1990). *Control of communicable diseases in man*. American Public Health Association, Washington, DC.

Bordewijk, J.A., Slaper, H., Reinen, H.A.J.M. & Schlamann, E. (1995). Total solar radiation and the influence of clouds and aerosols on the biologically effective UV. *Geophysical Research Letters*, 22, 2151-2154.

Bouma, M.J., Sondorp, H.E., & Kaay, H.J., van der (1994). Climate changes and periodic epidemic malaria. *Lancet*, 343, 1440.

Bowman, K.P. (1985). A global climatology of total ozone from the Nimbus-7 total ozone mapping spectrometer. In: Zerofos, C.S. & Ghazi, A. (eds). *Atmospheric ozone*. Reidel Publishing Co., Dordrecht, 363-367.

Box, E.O. (1981). *Macroclimate and plant forms: an introduction to predictive modelling in phytogeography*. Dr. J.W. Junk Publishers, The Hague.

Boyd, M.F. (1949). Epidemiology: factors related to the definitive host. In: Boyd, M.F. (ed). *Malariology (volume I)*. W.B. Saunders Company, Philadelphia and London, 608-697.

Bradley, D.J. (1993). Human tropical diseases in a changing environment. In: Ciba Foundation Symposium 175. *Environmental change and human health*. Wiley & Sons, Chichester, 147-170.

Brown, A.W.A. & Pal, R. (1971). *Insecticide resistance in arthropods*. World Health Organisation, Geneva.

Bull, G.M. (1980). The weather and deaths from pneumonia. *Lancet*, i, 1405-1408.

Bull, G.M. & Morton, J. (1975). Relationships of temperature with death rates from all causes and from certain respiratory and arteriosclerotic diseases in different age groups. *Age and Ageing*, 4, 232-246.

Bull, G.M. & Morton, J. (1978). Environment, temperature and death rates. *Age and Ageing*, 7, 210-224.

Carson, R. (1962). *Silent spring*. Penguin, London.

Carter, T.R., Parry, M.L., Harasawa, H. & Nishioka, S. (1994). *IPCC technical guidelines for assessing climate change impacts and adaptations*. IPCC/WMO/UNEP, Oxford.

CBS (1995). Population forecast for the Netherlands, 1995-2050 (in Dutch). *Statistisch Bulletin*, 51(50), 7-10.

CDC (1993). Schistosomiasis in US Peace Corps volunteers-Malawi 1992. *MMWR*, 42, 565-570.

Chandler, A.C. (1945). Factors influencing the uneven distribution of *Aedes aegypti* in Texas cities. *American Journal of Tropical Medicine and Hygiene*, 25, 145-149.

Charlson, R.J. & Wigley, T.M.L. (1994). Sulphate aerosol and climate change. *Scientific American*, 270(2), 48-57.

Christophers, S.R. (1960). *Aedes aegypti, the yellow fever mosquito.* Cambridge University Press, Cambridge.

Clements, A.N. & Paterson, G.D. (1981). The analysis of mortality and survival rate in wild populations of mosquitoes. *Journal of Applied Ecology*, 18, 373-399.

Collett, D. & Lye, M.S. (1987). Modelling the effect of intervention on the transmission of malaria in east Malaysia. *Statistics in Medicine*, 6, 853-861.

Coluzzi, M. (1994). Malaria and the Afrotropical ecosystems: impact of man-made environmental changes. *Parassitologia*, 36(1-2), 223-227.

Comins, H.N. (1977). The development of insecticide resistance in the presence of migration. *Journal of Theoretical Biology*, 64, 177-197.

Cubasch, U., Hasselmann, K., Hock, H., Maier-Reimer, E., Mikolajewicz, U., Santer, B.D. & Sausen, R. (1992). Time-dependent greenhouse warming computations with a coupled ocean-atmosphere model. *Climate Dynamics*, 8, 55-69.

Curtis, C.F., Cook, L.M. & Wood, R.J. (1978). Selection for and against insecticide resistance and possible methods of inhibiting the evolution of resistance in mosquitoes. *Ecological Entomology*, 3, 273-287.

Davidson, G. (1954). Estimation of the survival rate of anopheline mosquitoes in nature. *Nature*, 174, 792-793.

Detinova, T.S., Beklemishev, W.N. & Bertram, D.S. (1962). *Age-grouping methods in diptera of medical importance.* WHO Monograph 47, World Health Organisation, Geneva.

Dietz, K. (1988). Mathematical models for transmission and control of malaria. In: Wernsdorfer, W.H. & McGregor, I. (eds). *Malaria: principles and practice of malariology (volume 2).* Churchill Livingstone, New York, 1091-1133.

Doll, R. (1992). Health and the environment in the 1990s. *American Journal of Public Health*, 82(7), 933-941.

Donawho, C.K. & Kripke, M.L. (1991). Evidence that the local effect of ultraviolet radiation on the growth of murine melanomas is immunologically mediated. *Cancer Research*, 51, 4176-4181.

Driscoll, D.M. (1971). The relationship between weather and mortality in ten major metropolitan areas in the United States, 1962-1965. *International Journal of Biometeorology*, 15(1), 23-39.

Dunningan, M.G., Harland, W.A. & Fyfe, T. (1970). Seasonal incidence and mortality of ischaemic heart-disease. *Lancet*, 2, 793-796.

Dutch Health Council (1994). *UV radiation from sunlight* (in Dutch). Health Council of The Netherlands, The Hague, 1994 (05).

Dye, C.M. (1986). Vectorial capacity: must we measure all its components? *Parasitology Today*, 2, 203-209.

Dye, C.M. (1990). Epidemiological significance of vector-parasite interactions. *Parasitology*, 101, 409-415.

Dzidonu, C.K. & Foster, F.G. (1993). Prolegomena to OR modelling of the global environment-development problem. *Journal of the Operational Research Society*, 44(4), 321-331.

Elzen, M., den (1993). *Global environmental change: an integrated modelling approach.* PhD thesis, Maastricht University, International Books, Utrecht.

Engelen, G., White, R., Uljee, I. & Drazan, P. (1995). Using cellular automata for integrated modelling of socio-environmental systems. *Environmental Monitoring and Assessment*, 34, 203-214.

Ewald, P.W. (1994). *Evolution of infectious diseases.* Oxford University Press, Oxford.

Focks, D.A., Haile, D.G., Daniels, E., & Mount, G.A. (1993a). Dynamic life table model for *Aedes aegypti* (Diptera: Culicidae): analysis of the literature and model development. *Journal of Medical Entomology*, 30(6), 1003-1017.

Focks, D.A., Haile, D.G., Daniels, E. & Mount, G.A. (1993b). Dynamic life table model for *Aedes aegypti* (Diptera: Culicidae): simulation results and validation. *Journal of Medical Entomology*, 30(6), 1018-1028.

Focks, D.A., Daniels, E., Haile, D.G. & Keesling, J.E. (1995). A simulation model of the epidemiology of urban dengue fever: literature analysis, model development, preliminary validation, and samples of simulation results. *American Journal of Tropical Medicine and Hygiene*, 53(5), 489-506.

Forbes, P.D., Davies, R.E. & Urbach, F. (1978). Experimental ultraviolet photocarcinogenesis: Wavelength interactions and time-dose relationship. In: Kripke, M.L. & Sass, E.R (eds). *NCI monograph 50* NCI, Bethesda, MD, 31-38.

Freeman, T. (1995). *Malaria Zimbabwe 1995: a review of the epidemiology of malaria transmission and distribution in Zimbabwe and the relationship of malaria outbreaks to preceding meteorological conditions.* March 1995.

Freeman, T. & Bradley, M. (1996). Temperature is predictive of severe malaria years in Zimbabwe. *Transactions of the Royal Society of Tropical Medicine and Hygiene,* 90, 232.

Funtowicz, S.O. & Ravetz, J.R. (1989). *Managing uncertainty in policy-related research.* Paper presented at the International Colloquium 'Les experts sont formels: Controverses scientifiques et décisions politiques dans le domain de l'environment'. Arc et Sannes, France, 11-13 September 1989.

Garrett-Jones, C. (1964). The human blood index of malaria vectors in relation to epidemiological assessment. *Bulletin of the World Health Organisation*, 30, 241-261.

Garrett-Jones, C. & Grab, B. (1964). The assessment of insecticidal impact on the malaria mosquito's vectorial capacity, from the data on the populations of parous females. *Bulletin of the World Health Organisation*, 31, 71-86.

Garrett-Jones, C. & Shidrawi, G.R. (1969). Malaria vectorial capacity of a population of *Anopheles gambiae. Bulletin of the World Health Organisation*, 40, 531-545.

Garrett-Jones, C., Boreham, P.F.L. & Pant, C.P. (1980). Feeding habits of anophelines (Diptera: Cilcidae) in 1971-78, with reference to the human blood index: a review. *Bulletin of Entomological Research*, 70, 165-185.

Georghiou, G.P. & Taylor, C.E. (1977). Genetic and biological influences in the evolution of insecticide resistance. *Journal of Economic Entomology,* 70, 3, 319-323.

Giles, G., Marks, R. & Foley, P. (1988). Incidence of non-melanocytic skin cancer treated in Australia. *British Medical Journal*, 296, 13-17.

Gillies, M.T. (1988). Anopheline mosquitoes: vector behaviour and bionomics. In: Wernsdorfer, W.H. & McGregor, I. (eds). *Malaria: principles and practice of malariology (volume 1)*. Churchill Livingstone, New York, 453-486.

Gillies, M.T. & De Meillon, B. (1968). *The anophelinae of Africa south of the Sahara (Ethiopian zoogeographical region), 2nd edition.* South African Institute of Medical Research, Publication No. 54.

Glass, E.H., Adkinsson, P.L., Carlson, G.A. *et al.* (1984). *Pesticide resistance: strategies and tactics for management.* National Academy Press, Washington, DC.

Glass, G.E., Aron, J.L., Hugh Ellis, J. & Yoon, S.S. (1993). *Application of GIS technology to disease control.* Johns Hopkins University, Baltimore.

Goldberg, D. (1989). *Genetic algorithms in search, optimization, and machine learning.* Addison-Wesley, Reading, MA.

Granier, C. & Brasseur, G. (1992). Impact of heterogeneous chemistry on model predictions of ozone changes. *Journal of Geophysical Research*, 97, 18015-18033.

Green, M.S., Harari, G. & Kristal-Boneh, E. (1994). Excess winter mortality from ischaemic heart disease and stroke during colder and warmer years in Israel: An evaluation and review of the role of environmental temperature. *European Journal of Public Health*, 4, 3-11.

Gruijl, F.R., de & Leun, J.C., van der (1993). Influence of ozone depletion in the incidence of skin cancer. In: Young, A.R., Björn, L.O., Moan, J. & Nultsch, W. (eds). *Environmental UV-photobiology*. Plenum Press, New York and London, 89-112.

Gruijl, F.R., de & Leun, J.C., van der (1994). Estimate of the wavelength dependency of ultraviolet carcinogenesis in humans and its relevance to the risk assessment of stratospheric ozone depletion. *Health Physics*, 67(4), 319-325.

Gubler, D.J. (1976). Variation among geographic strains of *Aedes albopictus* in susceptibility to infection with dengue viruses. *American Journal of Tropical Medicine and Hygiene*, 25, 318-325.

Gubler, D.J. (1987). Current research on dengue. In: Harris, K.F. (ed). *Current topics in vector research, volume III.* Springer-Verlag, New York.

Gubler, D.J. & Clark, G.C. (1995). Dengue/dengue hemorrhagic fever: the emergence of a global health problem. *Emerging Infectious Diseases*, 1, 55-57.

Gunakasem, P., Chantrasri, C., Chaiyanum, S., Simasathien, P., Jatanasen, S. & Sangpetchsong, V. (1981). Surveillance of dengue hemorrhagic fever cases in Thailand. *Southeast Asian Journal of Tropical Medicine and Public Health*, 12, 338-343.

Haines, A. & Fuchs, C. (1991). Potential impacts on health of atmospheric change. *Journal of Public Health and Medicine*, 13(2), 69-80.

Halstead, S.B. (1993). Global epidemiology of dengue: health systems in disarray. *Tropical Medicine*, 35, 137-146.

Halstead, S.B. & Papaevangelou, G. (1980). Transmission of dengue 1 and 2 viruses in Greece in 1928. *American Journal of Tropical Medicine and Hygiene*, 29, 635-637.

Hamilton, A.C. (1989). In: Hamilton, A.C & Bensted-Smith, R. (eds). *Forest conservation in the East Usambaras*, IUCN, 97-133.

Harding, J.J. (1992). Physiology, biochemistry, pathogenesis, and epidemiology of cataract. *Current Opinion in Ophthalmology*, 3, 3-12.

Herrera-Basto, B.E., Prevots, D.R., Zarate, L., Luis, S.J. & Sepulveda, A.J. (1992). First reported outbreak of classical dengue fever at 1,700 meters above sea levels in Guerrero State, Mexico, June 1988. *American Journal of Tropical Medicine and Hygiene*, 46, 649-653.

Hill, D., White, V., Marks, R., Theobald, Th, Borland, R. & Roy, C. (1992). Melanoma prevention: behavioural and nonbehavioural factors in sunburn among an Australian urban population. *Preventive Medicine*, 21, 654-669.

Hofbauer, J. & Sigmund, K. (1988). *The theory of evolution and dynamical systems*. Cambridge University Press, Cambridge.

Holland, J.H. (1975). *Adaptation in natural and artificial systems*. University of Michigan Press, Ann Arbor, MI.

Holland, J.H. (1992). Genetic algorithms. *Scientific American*, July, 44-50.

Holman, C.J.D. & Armstrong, B.K. (1984). Cutaneous malignant melanoma and indicators of total accumulated exposure to the sun: an analysis separating histogenic types. *Journal of the National Cancer Institute*, 73, 75-82.

Horgan, J. (1995). From complexity to perplexity. *Scientific American*, June, 74-79.

Horsfall, W.R. (1955). *Mosquitoes: their bionomics and relation to disease*. Hafner Publishing Company, New York.

Hulme, M. (ed). (1996). *Climate change and southern Africa: an exploration of some potential impacts and implications in the SADC region*. WWF International, Climatic Research Unit, University of East Anglia, Norwich, UK.

Hunter, J.M., Rey, L., Chu, K.Y., Adekolu-John, E.O. & Mott, K.E. (1993). *Parasitic diseases in water resources development: the need for intersectoral negotiation*. World Health Organisation, Geneva.

Husain, Z., Pathak, M.A., Flotte, T. & Wick, M.M. (1991). Role of ultraviolet radiation in the induction of melanocytic tumours in hairless mice following 7,12-dimethylbenz(a)anthracene application and ultraviolet irradiation. *Cancer Research*, 51, 4964-4970.

IPCC (1990). *Climate change: the IPCC scientific assessment*. Houghton, J.T., Jenkins, G.J. & Ephraums, J.J. (eds). Cambridge University Press, New York.

IPCC (1991). *Climate change: the IPCC response strategies*. The Island Press, Washington DC.

IPCC (1996). *Climate change 1995: The science of climate change*. Houghton, J.J., Meiro Filho, L.G., Callander, B.A., Harris, N., Kattenberg, A. & Maskell, K. (eds). Cambridge University Press, New York.

Janssen, M.A. (1996). *Meeting targets: tools to support integrated assessment modelling of global change.* PhD thesis, Maastricht University.

Janssen, M.A. & Martens, W.J.M. (1996). *Managing malaria: an evolutionary modelling approach.* Globo Report Series No. 12, RIVM Report No. 461502012, Dutch National Institute of Public Health and the Environment, Bilthoven.

Janssen, M.A. & Martens, W.J.M. (1997). Modelling malaria as a complex adaptive system. *Artificial Life,* 3, 213-236.

Janssen, P.H.M., Slob, W. & Rotmans, J. (1990*). Sensitivity analysis and uncertainty analysis: an inventory of ideas, methods, and techniques* (in Dutch). National Institute of Public Health and the Environment, RIVM Report No. 958805001, Bilthoven.

Jetten, T.H. & Focks, D.A. (1997). Potential changes in the distribution of dengue transmission under climate warming. *American Journal of Tropical Medicine and Hygiene,* 57(3), 285-297.

Jetten, T.H., Martens, W.J.M. & Takken, W. (1996). Model simulations to estimate malaria risk under climate change. *Journal of Medical Entomology,* 33(3), 361-371.

Jetten, T.H. & Takken, W. (1994). *Anophelism without malaria in Europe: a review of the ecology and distribution of the genus Anopheles in Europe.* Agricultural University, Wageningen.

Jones, T.S., Liang, A.P., Kilbourne, E.M. *et al.* (1982). Morbidity & mortality associated with the July 1980 heatwave in St Louis and Kansas city, MO. *Journal of the American Medical Association,* 247 (24), 3327-3331.

Jordan, P. & Webbe, G. (1982). *Schistosomiasis: epidemiology, treatment and control.* William Heinemann, London.

Kalkstein, L.S. (1993). Health and climate change: direct impacts in cities. *Lancet,* 342, 1397-1399.

Kalkstein, L.S. & Davis, R.E. (1989). Weather and human mortality: an evaluation of demographic and interregional responses in the United States. *Annals of the Association of American Geographers,* 79(1), 44-64.

Kalkstein, L.S. & Smoyer, K.E. (1993). The impact of climate change on human health: some international implications. *Experientia,* 49, 969-979.

Katsouyanni, K., Pantazopoulou, A., Touloumi, G. *et al.* (1993). Evidence for interaction between air pollution and high temperature in the causation of excess mortality. *Archives of Environmental Health,* 48(4), 235-242.

Kauffman, S.A. (1991). Anti-chaos and adaptation. *Scientific American,* August, 64-70.

Keatinge, W.R., Coleshaw, S.R.K., Cotter, F. *et al.* (1984). Increases in platelet and red cell counts, blood viscosity, and arterial pressure during mild surface cooling: factors in mortality from coronary and cerebral thrombosis in winter. *British Medical Journal,* 289, 1405-1408.

Keatinge, W.R., Coleshaw, S.R.K. & Holmes, J. (1989). Changes in seasonal mortalities with improvement in home heating in England and Wales from 1964 to 1984. *International Journal of Biometeorology,* 33, 71-76.

Khan, A.Q. & Talibi, S.A. (1972). Epidemiological assessment of malaria transmission in an endemic area of East Pakistan and the significance of congenital immunity. *Bulletin of the World Health Organisation,* 46, 783-792.

Kilbourne, E.M. (1989). Heatwaves. In: Gregg, M.B. (ed). *The public health consequences of disasters.* US Department of Health and Human Services, Public Health Service, CDC, Atlanta, 51-61.

King, M. (1990). Health is a sustainable state. *Lancet,* 336, 664-667.

Kloos, H. & Thompson, K. (1979). Schistosomiasis in Africa: an ecological perspective. *The Journal of Tropical Geography,* 48, 31-46.

Kricker, A., Armstrong, B.K., English, D.R. & Heenan, P.J. (1995). Does intermittent sun exposure cause basal cell carcinomas? A control study in Western Australia. *International Journal of Cancer,* 60, 489-494.

Kricker, A., Armstrong, B.K. & McMichael, A.J. (1994). Skin cancer and ultraviolet radiation. *Nature,* 368(6472), 594.

Kunst, A.E., Looman, C.W.N. & Mackenbach, J.P. (1993). Outdoor air temperature and mortality in the Netherlands: a time-series analysis. *American Journal of Epidemiology,* 137(3), 331-341.

Langford, I.H. & Bentham, G. (1995). The potential effects of climate change on winter mortality in England and Wales. *International Journal of Biometeorology,* 38, 141-147.

Langton, C.G. (ed) (1989). *Artificial life.* Santa Fe Institute Studies in the Science of Complexity, Proceeding Part 6, Addison-Wesley, Redwood City, CA.

Larsen, U. (1990). The effects of monthly temperature fluctuations on mortality in the United States from 1921 to 1985. *International Journal of Biometeorology,* 34, 136-145.

Leach, J.F., McLeod, V.E., Pingstone, A.R., Davis, A. & Deane, G.W.H. (1978). Measurement of ultraviolet doses received by office workers. *Clinical and Experimental Dermatology,* 3, 77-79.

Leemans, R. & Cramer, W.P. (1991). *The IIASA database for mean monthly values of temperature, precipitation and cloudiness on a global terrestrial grid.* International Institute of Applied Systems Analysis, Laxenburg, Research Report RR-91-18.

Leeuw, F.A.A.M., de (1988): *Model based approach to the calculation of photolyze rates relevance for tropospheric chemistry* (in Dutch). RIVM Report No. 228603003, Dutch National Institute of Public Health and the Environment, Bilthoven.

Leeuw, F.A.A.M., de & Slaper, H. (1989): *UV radiation in the Netherlands: indication of the influence of a change in the ozone column* (in Dutch). RIVM Report No. 228903001, Dutch National Institute of Public Health and the Environment, Bilthoven.

Legates, D.R. & Willmott, C.J. (1990a). Mean seasonal and spatial variability in gauge-corrected global precipitation. *International Journal of Climatology,* 10(2), 111-127.

Legates, D.R. & Willmott, C.J. (1990b). Mean seasonal and spatial variability in global surface air temperature. *Theoretical and Applied Climatology,* 41(1-2), 11-21.

Leggett, J.A, Pepper, W.J. & Swart, R.J. (1992). Emissions scenarios for the IPCC: an update. In: IPCC. *Climate change 1992: the supplementary report to the IPCC scientific assessment.* Houghton, J.T., Callendar, B.A. & Varney, S.K. (eds), Cambridge University Press, New York, 69-95.

Levin, A.S, Hallam, T.G. & Gross, L.J. (eds) (1989). *Applied mathematical ecology.* Springer-Verlag, New York.

Levins, R. (1995). Toward an integrated epidemiology. *Trends in Ecology and Evolution,* 10(7), 304.

Levins, R., Albuquerque de Possas, C., Awerbuch, T. *et al.* (1993). *Preparing for new infectious diseases.* Harvard School of Public Health, Department of Population and International Health, Working Paper No. 8, June.

Levins, R., Awerbuch, T., Brinkman, U. *et al.* (1994). The emergence of new diseases. *American Scientist,* 82, 52-60.

Levy, S.B. (1992). *The antibiotic paradox: how miracle drugs are destroying the miracle.* Plenum Press, New York and London.

Lindsay, S. & Martens, W.J.M. (1998). Malaria in the African highlands: past, present and future. *Bulletin of the World Health Organisation,* in press.

Loevingsohn, M. (1994). Climatic warming and increased malaria incidence in Rwanda. *Lancet,* 343, 714-718.

Macdonald, G. (1957). *The epidemiology and control of malaria.* Oxford University Press, London.

Madronich, S. & Gruijl, F. R., de (1993). Skin cancer and UV radiation. *Nature,* 366(6450), 23.

Manabe, S., Stouffer, R.J., Spelman, M.J. & Bryan, K. (1991). Transient responses of a coupled ocean-atmosphere model to gradual changes of atmospheric CO_2. Part I: annual mean response. *Journal of Climate,* 4, 785-818.

Manabe, S., Spelman, M.J. & Stouffer, R.J. (1992). Transient responses of a coupled ocean-atmosphere model to gradual changes of atmospheric CO_2. Part II: seasonal response. *Journal of Climate,* 5(2), 105-126.

Marks, R., Staples, M. & Giles, G. (1993). Trends in non-melanocytic cancer treated in Australia: the second national survey. *International Journal of Cancer,* 53, 585-590.

Martens, W.J.M. (1996a). *Modelling the effect of global warming on the prevalence of schistosomiasis.* RIVM Report No. 461502010, GLOBO Report Series No. 10, Dutch National Institute of Public Health and the Environment, Bilthoven.

Martens, W.J.M. (1996b). Global atmospheric change and human health: an integrated modelling approach. *Climate Research,* 6, 107-113.

Martens, W.J.M (ed) (1996c). *Vulnerability of human population health to climate change: state-of-knowledge and future research directions.* Dutch National Research Programme on Global Air Pollution and Climate Change, Report No. 410200004 , Bilthoven.

Martens, W.J.M. (1997a). *Health impacts of climate change and ozone depletion: an eco-epidemiological modelling approach.* PhD thesis, Maastricht University.

Martens, W.J.M. (1997b). Climate change, thermal stress and mortality changes. *Social Science and Medicine,* 46(3), 331-344.

Martens, W.J.M., Rotmans, J. & Niessen, L.W. (1994). *Climate change and malaria risk: an integrated modelling approach.* RIVM Report No. 461502003, GLOBO Report Series No. 3, Dutch National Institute of Public Health and the Environment, Bilthoven.

Martens, W.J.M., Niessen, L.W., Rotmans, J., Jetten, T.H. & McMichael, A.J. (1995a). Potential impact of global climate change on malaria risk. *Environmental Health Perspectives,* 103(5), 458-464.

Martens, W.J.M., Jetten, T.H., Rotmans, J. & Niessen, L.W. (1995b). Climate change and vector-borne diseases: a global modelling perspective. *Global Environmental Change,* 5(3), 195-209.

Martens, W.J.M., Rotmans, J. & Vrieze, O.J. (1995c). Global atmospheric change and human health: more than merely adding up the risks. *World Resource Review,* 7(3), 404-416.

Martens, W.J.M, den Elzen, M.G.J., Slaper, H., Koken, P.J.M. & Willems, B.A.T. (1996). The impact of ozone depletion on skin cancer incidence: an assessment of the Netherlands and Australia. *Environmental Modelling and Assessment,* 1(4), 229-240.

Martens, W.J.M., Jetten, T.H., & Focks, D.A. (1997). Sensitivity of malaria, schistosomiasis and dengue to global warming. *Climatic Change,* 35, 145-156.

Martin, P.H. & Lefebvre, M.G. (1995). Malaria and climate: sensitivity of malaria potential transmission to climate. *Ambio,* 24(4), 200-207.

Mather, J.R. (1978). *The climatic water budget in environmental analysi*s. D.C. Heath and Company, MA.

Matsuoka, Y. & Kai, K. (1994). An estimation of climatic change effects on malaria. *Journal of Global Environment Engineering,* 1, 1-15.

McClelland, G.A.H. & Conway, G.R. (1971). Frequency of blood feeding in the mosquito *Aedes aegypti. Nature,* 232, 485-486.

McHugh, C.P. (1989). Ecology of a semi-isolated population of adult *Anopheles freeborni:* abundance, trophic status, parity, survivorship, gonotrophic cycle length, and host selection. *American Journal of Tropical Medicine and Hygiene,* 41 (2), 169-176.

McMichael, A.J. (1993). Global environmental change and human population health: a conceptual and scientific challenge for epidemiology. *International Journal of Epidemiology,* 22 (1), 1-8.

McMichael, A.J. (ed) (1996). Human population health. In: IPCC. *Climate change 1995: impacts, adaptations, and mitigation of climate change: scientific-technical analysis.* Watson, R.T, Zinyowera, M.C., Moss. R.H. and Dokken, D.J. (eds). Cambridge University Press, New York, 563-584.

McMichael, A.J. & Martens, W.J.M. (1995). The health impacts of global climate change: grappling with scenarios, predictive models and multiple uncertainties. *Ecosystem Health,* 1(1), 23-33.

McMichael, A.J., Haines, A., Slooff, R. & Kovats, S. (eds) (1996*). Climate change and human health; An assessment prepared by a Task Group on behalf of the World Health Organisation, the World Meteorological Organisation and the United Nations Environment Programme.* World Health Organisation, Geneva.

Meadows, D.H. & Robison, J.M. (1985). *The electronic oracle: computer models and social decisions.* John Wiley & Sons, Chichester.

Mitchell, J.F.B. & Ingram, W.J. (1992). Carbon dioxide and climate: mechanisms of changes in cloud. *Journal of Climate*, 5, 5-21.

Molineaux, L. (1988). The epidemiology of human malaria as an explanation of its distribution, including some implications for its control. In: Wernsdorfer, W.H. & McGregor, I. (eds). *Malaria: principles and practice of malariology (volume 2)*. Churchill Livingstone, New York, 913-998.

Molineaux, L. & Gramiccia, G. (1980). *The Garki Project: research on the epidemiology and control of malaria in the Sudan savannah of West Africa*. The World Health Organisation, Geneva.

Momiyama, M. & Katayama, K. (1972). Deseasonalization of mortality in the world. *International Journal of Biometeorology*, 16(4), 329-342.

Montzka, S.A., Butler, J.H., Myers, R.C. *et al*. (1996). Decline in the tropospheric abundance of halogen from halocarbons: implications for stratospheric ozone depletion. *Science*, 272(5266), 1318-1322.

Moore, C.G., Cline, B.L., Ruiz-Tibén, E., Lee, D., Romney-Joseph, H. & Rivera-Correa, E. (1978). *Aedes aegypti* in Puerto Rico: environmental determinants of larval abundance and relation to dengue virus transmission. *American Journal of Tropical Medicine and Hygiene*, 27, 1225-1231.

Morgan, G.M. & Henrion, M. (1990). *Uncertainty: a guide to dealing with uncertainty in quantitative risk and policy analysis*. Cambridge University Press, New York.

Morse, S.S. (1991). Emerging viruses: defining the rules for viral traffic. *Perspectives in Biology and Medicine*, 34(3), 387-409.

Muir, D.A. (1988). Anopheline mosquitoes: vector reproduction, life-cycle and biotope. In: Wernsdorfer, W.H. & McGregor, I. (eds). *Malaria: principles and practice of malariology (volume 1)*. Churchill Livingstone, New York, 431-452.

Murphy, J.M. (1995). Transient response of the Hadley Centre coupled ocean-atmospheric model to increasing carbon dioxide. Part I: control climate and flux correction. *Journal of Climate*, 8(1), 36-56.

Murphy, J.M. & Mitchell, J.F.B. (1995). Transient response of the Hadley Centre coupled ocean-atmospheric model to increasing carbon dioxide. Part II: spatial and temporal structure of response. *Journal of Climate*, 8(1), 57-80.

Murray, C.J.L. & Lopez, A.D. (1994). Global and regional cause-of-death patterns in 1990. *Bulletin of the World Health Organisation*, 72(3), 447-480.

Nájera, J.A. (1974). A critical review of the field application of a mathematical model of malaria eradication. *Bulletin of the World Health Organisation*, 50, 449-457.

Nájera, J.A., Liese, B.H. & Hammer, J. (1992). *Malaria: new patterns and perspectives*. World Bank Technical Paper Number 183, World Bank, Washington DC.

Näyhä, S. (1980). Short and medium-term variations in mortality in Finland. A study on cyclic variations, annual and weekly periods and certain irregular changes in mortality in Finland during the period 1868-1972. *Scandinavian Journal of Social Medicine*, Suppl. 21, 1-101.

Nedelman, J. (1985). Some new thoughts about some old malaria models. *Mathematical Biosciences*, 73, 159-182.

Neelemans, P.J. & Rampen, F.H.J. (1990). The epidemiology of the skin melanoma in the Netherlands (in Dutch). *Nederlands Tijdschrift Geneeskunde*, 135(42), 1238-1242.

Neering, H. & Cramer, M.J. (1988). Skin cancer in the Netherlands (in Dutch). *Nederlands Tijdschrift Geneeskunde*, 132, 1330-1333.

Newton, E.A.C. & Reiter, P.A. (1992). A model of the transmission of dengue fever with an evaluation of the impact of ultra-low volume, ULV, insecticide applications on dengue epidemics. *American Journal of Tropical Medicine and Hygiene*, 47, 709-720.

Oreskes, N., Shrader-Frechette, K. & Belitz, K. (1994). Verification, validation, and confirmation of numerical models in the Earth sciences. *Science*, 263, 641-646.

Pan, W.H., Li, L.A. & Tsai, M.J. (1995). Temperature extremes and mortality from coronary heart disease and cerebral infarction in elderly Chinese. *Lancet*, 345, 353-355.

Pant, C.P. (1988). Malaria vector control: imagociding. In: Wernsdorfer, W.H. & McGregor, I. (eds). *Malaria: principles and practice of malariology (volume 2)*. Churchill Livingstone, New York, 1173-1212.

Patz, J.A. & Balbus, J.M. (1996). Methods for assessing public health vulnerability to global climate change. *Climate Research*, 6, 113-125.

Patz, J.A., Epstein, P.R., Thomas, A.B., Burke, A. & Balbus, J.M. (1996). Global climate change and emerging infectious diseases. *The Journal of the American Medical Association*, 275, 217-223.

Patz, J.A., Martens, W.J.M, Focks, D.A., & Jetten, T.H. (1998). Simulation analysis of potential dengue fever risk under global climate change scenarios. *Environmental Health Perspectives*, 106(3), in press.

Paul, R.E.L., Packer, M.J., Walmsley, M., Lagog, M., Ranford-Carwright, L.C., Paru, R. & Day, K.P. (1995). Mating patterns in malaria parasite populations of Papua New Guinea. *Science*, 269, 1709-1711.

Peto, R., Lopez, A.D., Boreham, J., Thun, M. & Heath, C. (1994). *Mortality from smoking in developing countries 1950-2000*. Oxford University Press, New York.

Plüger, W. (1980). Experimental epidemiology of schistosomiasis. 1. The prepatent period and cercarial production of *Schistosoma mansoni* in *Biomphalaria* snails at various constant temperatures. *Parasitenkunde*, 63, 159-169.

Prah, S.K. & James, C. (1977). The influence of physical factors on the survival and infectivity of miracidia of *Schistosoma mansoni* and *S. haematobium* I. Effect of temperature and ultra-violet light. *Journal of Helminthology*, 51, 73-85.

Prather, M.J., Ibrahim, A.M.A., Sasaki, T., Stordal, F. & Visconti, G. (1992). Calculations of future Cl/Br and ozone depletion. In: *Scientific assessment of ozone depletion: 1991*. WMO/UNEP, WMO Global Ozone Research and Monitoring Project, Report No. 25, Chapter 8.

Pull, J.H. & Grab, B. (1974). A simple epidemiological model for evaluating the malaria inoculation rate and the risk of infection in infants. *Bulletin of the World Health Organisation*, 51, 507-515.

Purnell, R.E. (1966). Host-parasite relationships in schistosomiasis. I. The effect of temperature on the infection of *B. sud. tanganyiensis* with *S. mansoni* miracidia and of laboratory mice with *S. mansoni* cercariae. *Annals of Tropical Medicine and Parasitology*, 60, 90-96.

Ramaswamy, V., Schwarzkopf, M.D. & Shine, K.P. (1992). Radiative forcing of climate from halocarbon-induced global stratospheric ozone loss. *Nature*, 355, 810-812.

Risk Assessment Forum (1992). *Framework for ecological risk assessment*. Environmental Protection Agency, Washington DC.

Roberts, C.J. & Lloyd, S. (1972). Association between mortality from ischaemic heart disease and rainfall in south Wales and in the county boroughs of England and Wales. *Lancet*, 1, 1091-1093.

Rodin, E.Y. (ed) (1990). *Mathematical modelling: a tool for problem solving in engineering, physical, biological and social sciences*. Pergamon Press, Oxford.

Rogers, D.J. & Randolph, S.E. (1991). Mortality rates and population density of tsetse flies correlated with satellite imagery. *Nature*, 351, 739-741.

Rogot, E. (1974). Associations between coronary mortality and the weather, Chicago, 1967. *Public Health Reports*, 89(4), 330-338.

Rogot, E. & Blackwelder, W.C. (1970). Associations of cardiovascular mortality with weather in Memphis, Tennessee. *Public Health Reports*, 85(1), 25-39.

Rogot, E. & Padgett, S.J. (1976). Associations of coronary and stroke mortality with temperature and snowfall in selected areas of the United States, 1962-1966. *American Journal of Epidemiology*, 103(6), 565-575.

Root, T.L. & Schneider, S.H. (1995). Ecology and climate: research strategies and implications. *Science*, 269, 334-341.

Rose, G. (1966). Cold weather and ischaemic heart disease. *British Journal of Preventive and Social Medicine*, 20, 97-100.

Rosenzweig, C., Parry, M.L., Fisher, G. & Frohberg, K. (1993). *Climate change and world food supply*. Research Report No. 3, Environmental Change Unit, Oxford University, Oxford.

Ross, R. (1911). *The prevention of malaria* (2nd edition). Murray, London.

Rothman, K.J. (1993). Methodological frontiers in environmental epidemiology. *Environmental Health Perspectives Supplements*, 101 (4), 19-21.

Rotmans, J., Asselt, M.B.A., van, Bruin, A.J., de *et al.* (1994). *Global change and sustainable development: a modelling perspective for the next decade*. RIVM Report No. 461502004, GLOBO Report Series No. 4, Dutch National Institute of Public Health and the Environment, Bilthoven.

Rotmans, J., Dowlatabadi, H. & Parson, E.A. (1996). Integrated assessment of climate change: evaluation of models and other methods. In: Rayner, S. & Malone, E. (eds). *Human choice and climate change: an international social science assessment*. Cambridge University Press, New York.

Rotmans, J. & de Vries, H.J.M. (eds) (1997). *Perspectives on global futures: The TARGETS approach*. Cambridge University Press, Cambridge.

Rueda, L.M., Patel, K.J., Axtell, R.C. & Stinner, R.E. (1990). Temperature-dependent development and survival rates of *Culex quinquefasciatus* and *Aedes aegypti* (Diptera: Culicidae). *Journal of Medical Entomology*, 27, 892-898.

Rykiel, E.J., Jr. (1996). Testing ecological models: the meaning of validation. *Ecological Modelling*, 90, 224-229.

Sakamoto-Momiyama, M. (1978). Changes in the seasonality of human mortality: a medico-geographical study. *Social Science and Medicine*, 12, 29-42.

Sakamoto-Momiyama, M. & Katayama, K. (1971). Statistical analysis of seasonal variation in mortality. *Journal of the Meteorological Society of Japan*, 49(6), 494-508.

Schaanning, J., Finsen, H., Lereim, I. *et al.* (1986). Effects of cold air inhalation combined with prolonged sub-maximal exercise on airway function in healthy young males. *European Journal of Respiratory Diseases*, 68 (suppl. 143), 74-77.

Schapira, A. (1990). The resistance of *falciparum* malaria in Africa to 4-aminoquilines and antifolates. *Scandinavian Journal of Infectious Diseases*, Supplementum 75, 7-55.

Scott, T.W., Clark, C.G., Lorenz, L.H., Amerasinmghe, P.H., Reiter, P. & Edman, J.D. (1993a). Detection of multiple blood-feeding by *Aedes aegypti* during a single gonotrophic cycle using a histologic technique. *Journal of Medical Entomology*, 30, 94-99.

Scott, T.W, Chow, E., Strickman, D. *et al.* (1993b). Blood-feeding patterns of *Aedes aegypti* (Diptera: Culicidae) collected in a rural Thai village. *Journal of Medical Entomology*, 30, 922-927.

Scotto, J., Fears, T.R. & Fraumeni, J.F. (1981). *Incidence of non-melanoma skin cancer in the United States.* US Department of Health and Human Services, NIH 82-2433.

Service, M.W. (1965). Some basic entomological factors concerned with the transmission and control of malaria in northern Nigeria. *Transactions of the Royal Society of Tropical Medicine and Hygiene*, 59(3), 291-296.

Service, M.W. (1980). *A guide to medical entomology.* Macmillan, London.

Setlow, R.B., Grist, E., Thompson, K. & Woodhead, A.D. (1993). Wavelengths effective in induction of malignant melanoma. *Proceedings of the National Academy of Sciences*, 90, 6666-6670.

Sheppard, P.M., MacDonald, W.W., Tonn, R.J. & Grabs, B. (1969). The dynamics of an adult population of *Aedes aegypti* in relation to dengue haemorrhagic fever in Bangkok. *Journal of Animal Ecology*, 38, 661-702.

Shiff, C.J. (1964). Studies on *Bulinus (physopsis) globosus* in Rhodesia: the influence of temperature on the intrinsic rate of natural increase. *Annals of Tropical Medicine and Parasitology*, 58, 94-105.

Shiff, C.J., Coutts, W.C.C., Yiannakis, C. & Holmes, R.W. (1979). Seasonal patterns in the transmission of Schistosoma haematobium in Rhodesia, and its control by winter application of molluscicide. *Transactions of the Royal Society of Tropical Medicine and Hygiene*, 73, 375-380.

Shumway, R.H., Azari, A.S. & Pawitan, Y. (1988). Modelling mortality fluctuations in Los Angeles as functions of pollution and weather effects. *Environmental Research*, 45, 224-241.

Slaper, H., Schothorst, A.A., & Leun, J.C., van der (1986). Risk evaluation of UVBN therapy for psoriasis: comparison of calculated risk for UV-B therapy and observed risk in PUVA-treated patients. *Photodermatology*, 3, 271-283.

Slaper, H. (1987). *Skin cancer and UV exposure: investigations on the estimation of risks.* PhD thesis, University of Utrecht.

Slaper, H., Elzen, M.G.J., den, Woerd, H.J., van den & Greef, J., de (1992). *Ozone depletion and skin cancer incidence, an integrated modelling approach.* RIVM Report No. 249202, Dutch National Institute of Public Health and the Environment, Bilthoven.

Slaper, H., Velders, G.J.M., Daniel, J.S., Gruijl, F.R., de & Leun, J.C., van der (1996). Estimates of ozone depletion and skin cancer incidence to examine the Vienna Convention achievements. *Nature*, 384, 256-258.

Southwood, T.R.E. (1977). Habitat, the template for ecological strategies? *Journal of Animal Ecology,* 46, 337-365.

Southwood, T.R., Murdie, G., Yasuno, M., Tonn, R. J. & Reader, P.M. (1972). Studies on the life budget of *Aedes aegypti* in Wat Samphaya, Bangkok, Thailand. *Bulletin of the World Health Organisation*, 46(2), 211-226.

Stolarski, R.S., Bloomfield, P., McPeters, R.D. & Herman, J. (1991). Total ozone trends deduced from Nimbus 7 TOMS Data. *Geophysical Research Letters*, 18(6), 1015-1018.

Sturrock, B.M. (1973). Field studies on the transmission of *Schistosoma mansoni* and on the bionomics of its intermediate host, *Biomphalaria glabrata*, on St Lucia, West Indies. *International Journal for Parasitology*, 3, 175-194.

Susser, M. & Susser, E. (1996a). Choosing a future for epidemiology: I. Eras and paradigms. *American Journal of Public Health*, 86(5), 668-673.

Susser, M. & Susser, E. (1996b). Choosing a future for epidemiology: II. From black box to Chinese boxes and eco-epidemiology. *American Journal of Public Health*, 86(5), 674-677.

Tabashnik, B.E. & Croft, B.A. (1982). Managing pesticide resistance in crop-arthropod complexes: interactions between biological and operational factors. *Environmental Entomology*, 11, 1137-1144.

Tabashnik, B.E. (1990). Implications of gene amplification for evolution and management of insecticide resistance. *Journal of Economic Entomology,* 83(4), 1170-1176.

Taylor, C.E. (1983). Evolution of resistance to insecticides: the role of mathematical models and computer simulations. In: Georghiou, G.P. & Saito, T. (eds). *Pest resistance to pesticides.* Plenum, New York, 163-173.

Taylor, P. & Matambu, S.L. (1986). A review of the malaria situation in Zimbabwe with special reference to the period 1972-1981. *Transactions of the Royal Society of Tropical Medicine and Hygiene*, 80, 12-19.

The Eurowinter Group (1997). Cold exposure and winter mortality from ischaemic heart diseases, cerebrovascular diseases, respiratory disease, and all causes in warm and cold regions of Europe. *The Lancet*, 349, 1341-1346.

Thornthwaite, C.W. (1948). An approach towards a rational classification of climate. *The Geographical Review*, 38, 55-94.

UN (1992). *Long-range world population projections.* United Nations Population Division, New York.

UNEP (1991). *Environmental effects of ozone depletion: 1991 update.* UNEP, Nairobi.

UNEP (1994). *Environmental effects of ozone depletion: 1994 assessment.* UNEP, Nairobi.

Unganai, L.S. (1996). Historic and future climatic change in Zimbabwe. *Climate Research*, 6, 137-145.

Viner, D. (1994). Climate scenarios for impacts assessment. *Aspects of Applied Biology*, 38, 13-27.

Vitaliano, P.P. & Urbach, F. (1980). The relative importance of risk factors in nonmelanoma carcinoma. *Archives of Dermatology*, 116, 454-456.

Vitasa, B.C., Taylor, H.R., Strickland, P.T. *et al.* (1990). Association of nonmelanoma skin cancer and actinic keratosis with cumulative solar ultraviolet exposure in Maryland watermen. *Cancer*, 65, 2811-2817.

Von Bertalanffy, L. (1968). *General system theory.* New York.

Walsh, J.F., Molyneux, D.H. & Birley, M.H. (1993). Deforestation: effects on vector-borne diseases. *Parasitology*, 106, S55-75.

Watts, D.M., Burke, D.S., Harrison, B.A., Whitmire, R.E. & Nisalak, A. (1987). Effect of temperature on the vector efficiency of *Aedes aegypti* for dengue 2 virus. *American Society of Tropical Medicine and Hygiene*, 36(1), 143-152.

Weaver, W. (1948). Science and complexity. *American Scientist*, 36, 536-544.

Weidhaas, D.E., Breeland, S.G., Lofgren, C.S., Dame, D.A. & Kaiser, R. (1974). Release of chemosterilized males for the control of *Anopheles albimanus* in El Salvador. *The American Journal of Tropical Medicine and Hygiene*, 23 (2), 298-308.

Weil, K. & Kvale, K.M. (1985). Current research on geographical aspects of schistosomiasis. *The Geographical Review*, 75, 186-216.

Weinstock, M.A., Coldits, G.A., Willet, W.C., Stampfer, M.J., Rosner, B. & Speizer, F.E. (1991). Recall (report) bias and reliability in retrospective assessment of melanoma risk. *American Journal of Epidemiology*, 133, 240-245.

Wernsdorfer, W.H. (1994). Epidemiology of drug resistance in malaria. *Acta Tropica*, 56, 143-156.

West, R.R & Lowe, C.R. (1976). Mortality from ischaemic heart disease: inter-town variation and its association with climate in England and Wales. *International Journal of Epidemiology*, 5(2), 195-201.

White, G.B. (1982). Malaria vector ecology and genetics. *British Medical Bulletin*, 38, 207-212.

WHO (1990). *Potential health effects of climatic change.* WHO, Geneva.

WHO (1992). *Vector resistance to pesticides.* Fifteenth Report of the WHO Expert Committee on Vector Biology and Control, WHO Technical Report Series, 818, Geneva.

WHO (1993a). *The control of schistosomiasis.* WHO Technical Report Series, 830, Geneva.

WHO (1993b). *World health statistics annual.* WHO, Geneva.

WHO (1994). *Ultraviolet radiation: an authoritative scientific review of environmental and health effects of UV, with reference to global ozone layer depletion.* Environmental Health Criteria 160, WHO, Geneva.

WHO (1995a). *World health report 1995*. WHO, Geneva.

WHO (1995b). *Epidemiology and prevention of cardiovascular diseases in elderly people.* WHO Technical Report Series 853, WHO, Geneva.

WHO (1995c*). Key issues in dengue vector control toward the operationalization of a global strategy.* WHO Report CTD/FL(DEN)/IC/96.1, WHO, Geneva.

WHO (1996). *The world health report 1996: fighting disease fostering development*. WHO, Geneva.

Willems, B.A.T. & Koken, P.J.M. (1995). *Red on the greens? Modelling the ozone depletion, UV radiation and skin cancer rates for Australia.* Master's thesis, Maastricht University.

Winter-Sorkina, R., de (1995). *Depletion and natural variability of the ozone layer from TOMS observations.* RIVM Report No. 722201005, Dutch National Institute of Public Health and the Environment, Bilthoven.

WMO (1992). *Scientific assessment of ozone depletion: 1991*. WMO/UNEP, WMO Global Ozone Research and Monitoring Project, Report No. 25.

Woolhouse, M.E.J. & Chandiwana, S.K. (1990). Population dynamics model for *Bulinus globosus*, intermediate host for *Schistosoma haematobium*, in river habitats. *Acta Tropica,* 47, 151-160.

World Bank (1993). *World development report 1993; investing in health*. Oxford University Press, New York.

Zahar, A.R. (1974). Review of the ecology of malaria vectors in the WHO Eastern Mediterranean region. *Bulletin of the World Health Organisation*, 50, 427-440.

Zoysa, A.P.K., de, Herath, P.R.J., Abhayawardana, T.A., Padmalal, U.K.G.K. & Mendis, K.N. (1988). Modulation of human malaria transmission by anti-gamete blocking immunity. *Transactions of the Royal Society of Tropical Medicine and Hygiene*, 82, 548-553.

Zulueta, J., de (1994). Malaria and ecosystems: from prehistory to posteradication. *Parassitologia*, 36(1-2), 7-15.

Zulueta, J., de, Ramsdale, C.D. & Coluzzi, M. (1975). Receptivity to malaria in Europe. *Bulletin of the World Health Organisation*, 52, 109-111.

INDEX